D0906451

ROBERT E. KRIEGER
PUBLISHING COMPANY INC.
MALABAR
FLORIDA 32950

THE APPLICATIONS OF HOLOGRAPHY

WILEY SERIES IN PURE AND APPLIED OPTICS

Advisory Editor
STANLEY S. BALLARD University of Florida

Lasers, BELA A. LENGYEL
Ultraviolet Radiation, second edition, LEWIS R. KOLLER
Introduction to Laser Physics, BELA A. LENGYEL
Laser Receivers, MONTE ROSS
The Middle Ultraviolet: its Science and Technology, A. E. S. GREEN, *Editor*
Optical Lasers in Electronics, EARL L. STEELE
Applied Optics. A Guide to Optical System Design/Volume 1, LEO LEVI
Laser Parameter Measurements Handbook, HARRY G. HEARD
Gas Lasers, ARNOLD L. BLOOM
Advanced Optical Techniques, A. C. S. VAN HEEL, *Editor*
Infrared System Engineering, R. D. HUDSON
Laser Communication Systems, WILLIAM K. PRATT
Optical Data Processing, A. R. SHULMAN
The Applications of Holography, H. J. CAULFIELD and SUN LU

The Applications of Holography

H. J. CAULFIELD

Sperry Rand Research Center
Sudbury, Massachusetts

SUN LU

Texas Instruments Incorporated
Dallas, Texas

WILEY-INTERSCIENCE,
A DIVISION OF JOHN WILEY & SONS

NEW YORK · LONDON · SYDNEY · TORONTO

Library of Congress Catalogue Card Number: 77-107585

SBN 471 14080 5

Printed in the United States of America

10 9 8 7 6 5 4 3 2

Preface

With holography, we can make highly realistic three-dimensional pictures. We can study dynamic effects at our leisure. We can view the effects of arbitrary stresses on objects. We can read words by machine. We can make masks for integrated circuit manufacturing. We can see sound. We can do so many things that are possible in no other way.

The purpose of this book is to make the right information accessible to the right people. Holography is a delightful art practiced by a group of specialists who call themselves "holographers." It was our intention to write a book not for our fellow holographers, but for the many technical people who could use holography if they knew what it can do and how it can be done. Therefore, we have tried to make the book a self-contained introduction to holography and its applications. No effort is made to achieve mathematical elegence. Rather, the emphasis is on making the basic phenomena as clear as possible.

Obviously, a short, self-contained book on a very complex subject must either omit or, at best, merely allude to much of the detail. We describe just enough holography to give the reader a good idea of whether he wants to use it. If he does, he must acquire elsewhere some of the details not included in this book. However, even this reader should not throw the book away—it can still be useful! It can serve as a guide to further reading and to the language of holographers. It ceases to be a source book and becomes a handbook.

At this point, we must confess to second motive in writing the book. We wish to try to show others what we find to be so fascinating about holography. It has power and beauty, which have produced many holographers and few former holographers. Holography is fun. We hope the reader will be able to sense this as he follows the development in this book.

Even a book that is small and simple in concept takes a great deal of effort. Much of that effort is made by people whose names appear only at the end of the preface, and to them must go much of the credit and little of the blame. For support and encouragement, we thank Dr. R. W. Damon, Dr. A. R.

Hilton, Dr. W. J. Beyen, and Dr. H. W. Hemstreet. For their contributions and patience, we thank our wives. For such help as phrase correction, spelling correction, and typing, we thank one of those patient wives—Mrs. H. J. Caulfield.

H. J. Caulfield
Sun Lu

Sudbury, Massachusetts
Dallas, Texas
November 1969

Contents

XIV. APPLICATIONS OF HOLOGRAPHY TO MICROSCOPY

XV. APPLICATIONS OF HOLOGRAPHY TO MOTION PICTURES AND TELEVISION

XVI. THE FUTURE OF HOLOGRAPHY

THE APPLICATIONS OF HOLOGRAPHY

I

Waves and Images

A. THE WAVEFRONT RECONSTRUCTION PROBLEM

Most of the information reaching us comes to us in the form of waves—sound and especially light. A unique feature of our perception of wave information is that we perceive an object as being " out there." It appears to be located at a distinct place in space. Holography is a means of storing this wave information and regenerating it in such a way that no information is lost in the process. The reconstruction of the original wave information allows the observer to perceive the original object " out there " in three dimensions. The wavefront reconstruction problem can be stated simply: " How can we store the wave information in such a way that we can reconstruct the original wave information at will and allow our sense organs and nervous system to perceive the original objects?" To make the problem clearer, we must define the term "wavefront." Hereafter "wavefront" will mean the instantaneous state of the waves on some two-dimensional surface.* To create a wavefront, we freeze the waves in time and sample them on some two-dimensional surface.

Wavefronts are important because all the three-dimensional information about the waves is contained in the two-dimensional sample. Waves forced to conform to the original waves on a two-dimensional surface will thereafter duplicate the behavior of the original waves. Thus, no three-dimensional information is lost in the two-dimensional sampling. This remarkable property of waves is called the "boundary-value" property. The mathematical statement of the boundary-value problem will be found in Section II-C. Wavefront reconstruction reproduces the wave as a whole. Thus, the problem of recreating a three-dimensional "time-frozen" copy of the waves is reduced to that of reconstructing the wavefront on any two-dimensional surface.

An important qualification must now be stated. If the two-dimensional

---

* The word "wavefront" is used in other contexts to mean a uniphase surface. The usage in this book is more in line with the usual terminology of holography.

1

surface is a finite surface, some of the original wavefront information is neither recorded nor reconstructed. There are an infinite number of possible wavefronts which have the same value on the finite two-dimensional surface. The reconstruction process cannot resupply the lost information. Therefore, the three-dimensional wave reconstructed from a finite two-dimensional sample of the wavefront cannot be an exact copy of the original. In fact, the reconstructed wave is the sum of all of the waves which have the same wavefront at the finite two-dimensional sampling surface.* This uncertainty in the reconstructed wave is intimately connected with the Heisenberg uncertainty principle. Specifying that a photon passes through a two-dimensional surface of width Δx restricts our knowledge of the transverse momentum, p_x, to an uncertainty, Δp_x, such that

$$\Delta x \, \Delta p_x \approx h, \qquad (I\text{-}1)$$

where h is Planck's constant. The fractional uncertainty in momentum is the transverse momentum uncertainty, Δp_x, divided by the longitudinal momentum, p_0. Thus,

$$\frac{\Delta p_x}{p_0} \approx \frac{h/\Delta x}{h/\lambda} = \frac{\lambda}{\Delta x}, \qquad (I\text{-}2)$$

where the de Broglie relation

$$p_0 = \frac{h}{\lambda} \qquad (I\text{-}3)$$

has been used and λ is the wavelength. If we interpret momentum in terms of mass times velocity, the transverse velocity divided by the longitudinal velocity varies between zero and $\Delta p_x/p_0$. This represents an angular uncertainty of

$$\Delta\theta \approx \frac{\lambda}{\Delta x}. \qquad (I\text{-}4)$$

This is the "diffraction-limited" resolution of an aperture of width Δx. Further aspects of this problem can be found in the literature.[1] For the present purposes, it suffices to note that the larger the two-dimensional surface, the more accurate the three-dimensional reconstruction. Section II-C discusses diffraction limitations in mathematical terms.

---

* A more precise statement is that the hologram reconstructs not the original wavefront, but a wavefront identical to the original, on the sampling surface and identically zero elsewhere.

B. EARLY FORMS OF WAVEFRONT RECONSTRUCTION

It is only partially fair to review all previous wave-recording and wave-disturbing mechanisms as unsuccessful attempts at wavefront reconstruction. Nevertheless, a review of those mechanisms in terms of wavefront reconstruction will help to point out the problems associated with wavefront reconstruction. We will review intensity-only mechanisms, phase-only mechanisms, and, finally, combination mechanisms.

The intensity of a wave is a measure of the power the wave can deliver. If we record the intensity of a wave at each point on a two-dimensional surface and do not record the direction of propagation of the wave at each point, we have not recorded the wavefront. The absence of directional information means that the intensity-only recording cannot contain information about the wavefront some distance from the recording medium. The information on wavefront propagation is lost. Photography " solves " this problem in a very clever way. It uses a lens to produce a wavefront at the photographic medium which is a copy of the wavefront at the remote object. Without a lens, photography would record the intensity distribution of the wavefront from the remote object. It is easy to confirm that such a photograph is a featureless blob and is devoid of information about the object. One of the motives in the use of holography is to avoid the necessity of a lens (lenses for electrons, phonons, X-rays, etc., are very difficult to construct). The idea of holography (often called " lensless photography ") is to record the full wavefront, so the wavefront at any point in space can be examined.

A monochromatic wave which arises at some point in space due to a stationary object undergoes amplitude variations of the form

$$U = U_0 \cos (\omega t + \phi), \tag{I-5}$$

where U is the wave vector, U_0 is the amplitude of U, ω is the angular frequency, and ϕ is a scalar called the phase. Two waves differing in phase by any value other than integral multiples of $2\pi/r$ each comparable states, say maxima, at different times. As we will see in Section II-A, ϕ is related to the direction of propagation. Therefore, recording ϕ at every point in the wavefront is equivalent to recording the direction of propagation at every point.

Whereas the intensity of the wave (proportional to $U_0{}^2$) was recorded directly by photography, the phase was not recorded directly prior to holography. On the other hand, phase manipulations by phase plates, lenses, mirrors, etc. are common. A lens changes the phase (direction) of a ray without changing its intensity. Figure I-1 shows how a lens can be used for producing a real or a virtual image of a point source. In Figure I-1a we note that a point source emits a spherical wave pattern which the human, intercepting part of that wave pattern with his eyes, will see as a point in space. Any

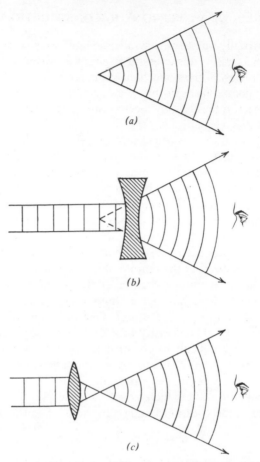

Figure I-1 Three cases in which a human viewer will see a point in space. (*a*) An actual point light source is present. The human, intercepting part of the diverging spherical wavefront with his eyes, sees the point at its actual location. (*b*) A collimated light source is diverged to form a virtual image of a point. (*c*) A collimated light source is converged to form a real image. In all three cases identical wavefronts reach the eye.

device which produces that same wave pattern at the observer's eyes will lead to his perception of that same point. Two ways of using lenses to convert a plane wave into a spherical wave are also shown in Figure I-1. In Figure I-1*b*, the point which the human perceives by extrapolation from the data reaching his eyes is not a point which is physically present. For example, a photographic plate placed coincident with the perceived point would not record an image of a point. For this reason, the image seen by the human viewer is said to be a "virtual" image. Figure I-1*c* shows how another kind of lens can be used to produce a real, physically present point and, hence, a "real" image. The distinction between real and virtual images is of great importance in holography.

Some attempts have been made to record both phase and intensity prior to holography. These methods can be called stereo methods. For example, two photographs of the same object viewed from separate places in space can be so encoded that the left eye sees one image and the right eye sees the other. The reaction of the majority of humans is to attempt to "fuse" the two-dimensional pictures into a single three-dimensional image. The differences between the two images are thus interpreted as depth cues. The trouble is that the directional phase information is fairly accurate only if the two images are viewed from that particular place in space which corresponds to the camera positions used to produce those images. The same applies to sound stereo systems. They sound right only if the listner is seated at the proper place relative to the two speakers. To overcome this limitation, the stereo process can be extended to include many images from many "points of view." This process, invented by Lippmann,[2] has been called integral photography. Lippmann integral photography, using a fly's eye lens, can produce images which have the correct phase information when viewed from many positions. The image appears to be obscured by a screen (due to the fly's eye lens), but holography can be used to eliminate this effect (Section VIII-G). At this point we are treading on the borderline of holography.

C. THE HISTORICAL DEVELOPMENT OF HOLOGRAPHY

The term "holography" was coined by the inventor of holography, Dennis Gabor.[3] It comes from the Greek "holos" meaning "the whole or entirely." A hologram records the whole of the wave information (U_0 and ϕ from Eq. I-5) about the wave at each point on a two-dimensional surface. Furthermore, the wavefront is easily reconstructed, so the whole three-dimensional information about the wave can be retrieved at will.

Gabor's first paper in 1948[3] introduced the essential concept for holographic recording—the reference beam. The use of the reference beam is necessitated by the fact that the physical detectors and recorders are sensitive only to the intensity $|U|^2 = U_0^2$. The phase, ϕ, is not recorded. In fact, the phase is manifest only when two waves of the same frequency are simultaneously present at the same place and if they are coherent (Section II-B). In that case the two waves combine to form a single wave whose intensity depends not only on the intensities of the two individual waves, but also on the phase difference between them. This is the key to holography.

We will illustrate holography with the special case of a hologram of waves focused to a point. The wave information, $U(x)$, in the plane of the point is

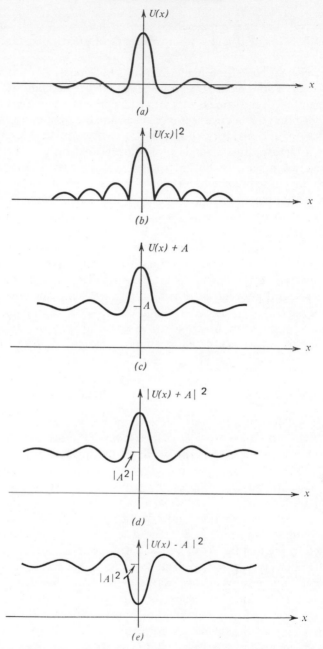

Figure I-2 Recording the wavefront as a result of a focused source: (*a*) the actual wavefront, $U(x)$; (*b*) the square-law detected wavefront, $|U(x)|^2$; (*c*) $U(x)$ plus a strong coherent reference wave A; (*d*) the square-law detected wavefront, $|U(x) + A|^2$; and (*e*) the spare-law detected wavefront, $|U(x) - A|^2$.

shown in Figure I-2a. The albegraic sign of $U(x)$ is used to specify the phase information. The "+ regions" and the "− regions" differ in phase by π (one-half wavelength). A photographic plate in the focal plane would record the pattern $|U(x)|^2$ shown in Figure I-2b. If a uniform, coherent reference wave A_0 were also present, then the total amplitude would be

$$U_T = A_0 + U(x)$$

as shown in Figure I-2c. A physical recording medium would record $|U_T|^2$ as shown in Figure I-2d. It would be possible to shift the phase of the reference beam by π to achieve a recording as shown in Figure II-2e. Whereas Figure I-2b (no reference beam) shows peaks for the peaks of both phases of the original diffraction pattern (Figure I-2a), the recorded patterns (Figures I-2d and I-2e) with a reference beam show only those peaks which are in phase with the reference beam. The two recordings between them contain all the phase and intensity information.[4] One contains the + phase information and the other contains the − phase information. The progress in holography, has been made because of a surprising fact. Either of the recordings of Figures I-2d and I-2e can be used to reconstruct the original wavefront! This is actually a result from classical optics called Babinet's principle.* The fact that information has been lost (when we, effectively, record one phase only) leads to ambiguity in the reconstructed wavefront. In fact, either of two "conjugate" wavefronts could produce the hologram. In the reconstruction, nature "solves" this problem by producing both wavefronts. The two images arising from this ambiguity are different from each other. Normally one is a real image and the other is a virtual image. Much of the early development of holography was aimed at reducing the effect of the extra image.[6-9]

At this point we must describe Gabor's original experiments. (For details of this work see Section IV-B.) His light source was filtered to make it highly monochromatic and passed through a small pinhole to assure that good interference patterns could be formed between various parts of the diverging beam. The object was a small transparency of alpha-numeric characters. The light diffracted by the alpha-numeric characters and the light passing through and around the transparent region formed the object and reference beams which produced an interference pattern at the photographic plate. The developed photographic plate (a positive of the interference pattern) was the first hologram. When illuminated by a monochromatic copy of the diverging beam, the hologram caused come of the light to be diffracted to form images of the original object.

* Babinet's principle states that complimentary screens produce identical Fraunhofer diffraction patterns. This is true only under certain highly restrictive conditions.[5] The invariance of the holographic wavefront with the sign of the photographic gamma can be considered to be an exact form of Babinet's principle.

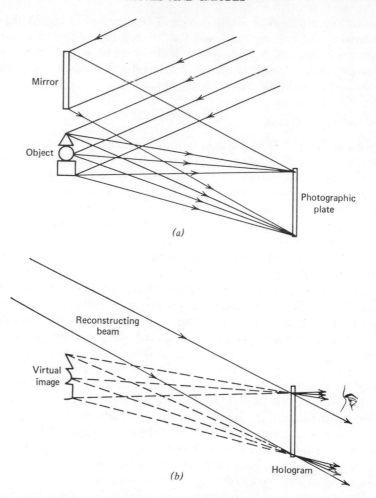

Figure I-3 The split-beam hologram method owing to Leith and Upatnieks: (*a*) a typical hologram formation setup and (*b*) reconstruction of the object wavefront using the hologram from (*a*).

The real and virtual images were both on the optical axis. This led to two viewing difficulties. First, both images were obscured by the undiffracted portion of the reconstructing beam. Second, each image obscured the other. A camera focused on either image would also record the out-of-focus image of the other.

The first report of a breakthrough eliminating both of those image-degrading difficulties was a paper by Leith and Upatnieks in 1962.[10] The time lapse from Gabor's paper in 1948 to the Leith-Upatnieks paper in 1962 was not caused by a lack of interest or cleverness among holographers. Rather, it seems to have been caused by the lack of a bright, highly coherent light source. The laser provided just such a source. The Leith-Upatnieks paper provided a technique for using laser light to improve hologram imaging. The most important feature of the Leith-Upatnieks paper was the demonstration that with lasers it was no longer necessary to have both the reference beam and the object beam incident from the same direction. The Leith-Upatnieks split-beam method is sketched in Figure I-3*a*. The hologram can be reconstructed* as shown in Figure I-3*b*. Note that the two images and the undiffracted beam can now be physically separated. This permits highly realistic, three-dimensional photography.

Since 1962 interest in holography has become widespread and intense. Among the developments since then are many of practical utility, these are discussed in Chapters VIII through XV. It is already clear that new applications of holography are emerging. The future of holography (Chapter XVI) will be written not just by the holographers, but also by the users of holography This book was written for them.

REFERENCES

1. H. J. Caulfield, *Amer. J. Phys.* **34**, 1066 (1966).
2. G. Lippmann, *Compt. Rend.*, **146**, 446 (1908), *J. Phys.*, **7**, 821 (1908).
3. D. Gabor, *Nature*, **161**, 777 (1948); "Microscopy by Reconstructed Wave-fronts," *Proc. Roy. Soc. (London)*, **A197**, 454 (1947).
4. D. Gabor and W. P. Gross, *J. Opt. Soc. Amer.*, **56**, 849 (1966).
5. H. Lipson and K. Walkley, *Opt. Acta*, **15**, 83 (1968).
6. D. Gabor, *Proc. Phys. Soc. (London)*, **B64**, 449 (1951).
7. A. V. Baez, *J. Opt. Soc. Amer.* **42**, 756 (1952).
8. A. V. Baez and H. M. A. El-Sum, "Effects of Finite Source Size, Radiation Bandwidth, and Object Transmission in Microscopy by Reconstructed Wavefronts" in *X-Ray Microscopy and Microradiology Proceedings*, Eds., V. S. Cosslett, A. Engstrom, and H. H. Pattee, Jr., Academic Press, New York, 1957 p. 347.
9. A. Lohmann, *Opt. Acta*, **3**, 97 (1956).
10. E. N. Leith and J. Upatnieks, *J. Opt. Soc. Amer.* **52**, 1123 (1962).

---

* Literally, the hologram is not "reconstructed"; it allows reconstruction of the object wavefront. The term "reconstructing the hologram" is a convenient and very common shorthand which means "reconstructing the object wavefront using the hologram." No confusion can result from such a usage if the meaning of the shorthand is understood.

II

Waves—Their Properties and Sources

A. UNIFYING CONCEPTS

Holography is a means for studying waves—any waves. No particular type of wave is required. To define the subject matter of holography we must define the word "wave". A wave is something which can be represented by a scalar function $U(\mathbf{x}, t)$ satisfying the equation

$$\nabla^2 U - \frac{1}{c^2} \frac{\partial^2 U}{\partial t^2} = 0, \qquad \text{(II-1)}$$

where

$$\nabla^2 = (\partial^2/\partial x^2) + (\partial^2/\partial y^2) + (\partial^2/\partial z^2),$$

$\mathbf{x} = x\hat{\imath} + y\hat{\jmath} + z\hat{k}$ is the space coordinate vector, t is the time coordinate, and c is a positive, real scalar called the wave-propagating speed. If we restrict ourselves to harmonic solutions to the wave equation,

$$U = A(x)e^{i\omega t}, \qquad \text{(II-2)}$$

then

$$\nabla^2 U = e^{i\omega t} \nabla^2 A,$$

$$\frac{\partial^2 U}{\partial t^2} = -\omega^2 A e^{i\omega t},$$

and, therefore,

$$\nabla^2 A + k^2 A = 0, \qquad \text{(II-3)}$$

where $k = (\omega/c)$. The general solution of Eq. II-3 is

$$A = A_1 \exp{(i\mathbf{k} \cdot \mathbf{x})} + A_2 \exp{(-i\mathbf{k} \cdot \mathbf{x})},$$

where $k\hat{\mathbf{r}} = \mathbf{k}$ and $\hat{\mathbf{r}}$ is the unit vector in the wave-propagation direction. Thus we have

$$U = A_1 \exp{i(\mathbf{k} \cdot \mathbf{x} + \omega t)} + A_2 \exp{i(-\mathbf{k} \cdot \mathbf{x} + \omega t)}. \qquad \text{(II-4)}$$

B. WAVE PROPAGATION

The wave equation has a very significant property, called linearity, which is easily explained. If two waves, U_1 and U_2, satisfy the wave equation, then $U = U_1 + U_2$ also satisfies the wave equation. The theory of linear equations is well tabulated in mathematic a texts,[1] so we can immediately use it to describe wave propogation.

Of primary importance is the principle of superposition, which follows from the linearity property. According to this principle, a wave, U, can be considered as the sum of individual, independently propogating waves, U_i, providing only that $U = \sum U_i$.

We can now derive the boundary value property which was noted in Section-I-A as being the basis for holographic wavefront reconstruction. Consider an infinite, two-dimensional surface S. We divide S into small regions S_i. If the wavefront on each region is denoted by O_i, the wavefront on S is $O = \sum O_i$. The question is: "What will be the wavefront, W, at a point x_0 not on the original surface S?" The answer is

$$W = \sum W_i, \tag{II-5}$$

where W_i is the wavefront at x_0 due to O_i. To evaluate W properly, we must take into account phase shifts due to the fact that different S_i's are at different distances from x_0. In order to define those phase shifts accurately, we allow the S_i's to become very small in which case each S_i is essentially a point source. We know that the amplitude of the wave from a point source decreases inversely as distance r from the source to the point of interest, x_0. Furthermore, phase varies as e^{ikr}, so the contribution due to O_i is proportional to

$$W_i = \frac{A_i S_i e^{ikr}}{r} \tag{II-6}$$

where A_i is the average value of $|O_i|$ over S_i. In this limit, Eq. (II-5) becomes

$$W(x_0) = \int_S \frac{A(\mathbf{x}) e^{ikr(\mathbf{x}, \, x_0)}}{|\mathbf{x} - \mathbf{x}_0|} \, d\mathbf{x}. \tag{II-7}$$

Thus, a detailed knowledge of the wavefront on S is sufficient to determine the wavefront at x_0 not on S. This being true, we can create the proper wavefront at x_0 by forcing a wave to conform to O on the surface S. This is exactly what holography does.

A very important special case of Eq. II-7 occurs if we consider an aperture in a plane opaque screen, S under the following conditions:

1. The distance r_1 from the source to an arbitrary point in the aperture is very large compared with the aperture dimensions.

2. The distance, r_2, from an arbitrary point in the aperture to any particular point, P, in the diffraction pattern is large compared with the aperture dimensions.

3. The aperture dimensions are large compared with the wavelength.

The latter condition allows us to make accurate calculations on the basis of a very simple assumption on the boundary conditions at S. We assume that the wavefront due to the source is undisturbed by the screen at all positions within the aperture and zero everywhere else. If all these conditions hold, the process is called Fraunhofer diffraction, and the expression for the wavefront in the x–y plane is

$$W(x, y) = C \iint\limits_{-\infty}^{\infty} T(\beta, \gamma) \exp\left[-i(\beta x + \gamma y)\, d\beta\, d\gamma\right], \tag{II-8}$$

where $\beta = (ku/r)$, $\gamma = (kv/r)$, u and v are the coordinates in the aperture, $T(\beta, \gamma)$ is the transmission of the aperture at coordinates (u, v), and C is a constant. Note that $W(x, y)$, except for a multiplicative constant, is a Fourier transform of the transmission. If the conditions for Fraunhofer diffraction do not hold, there is no single general version of Eq. II-7 (although several special cases can be treated in closed form). In this case we speak of Fresnel diffraction. The relation of Fresnel and Fraunhofer diffraction to holography is discussed in Section VI-E.

Fraunhofer diffraction provides a convenient way to discuss imaging with lenses. An ideal lens is designed to take a bundle of parallel rays and converge them to a single spot on the focal plane. (The Fourier transform of a plane wave is a delta function.) Likewise, a lens is designed to transform a point source on the focal plane into a beam of parallel rays. (The Fourier transform of a delta function is a plane wave.) Yet another way of relating the two focal planes of a lens is to say that angles in one plane correspond to positions in the other plane and vice versa. The most meaningful way to express all these observations is to say that, except for a multiplicative constant which we omit, a wave $f(x, y)$ in one focal plane is Fourier transformed into a wave

$$F(u, v) = \int_{-\infty}^{\infty} f(x, y) e^{i(xu + yv)}\, dx\, dy, \tag{II-9}$$

where the spatial coordinates ξ and η in the second plane are given by

$$\xi = \frac{fu}{k}$$

and

$$\eta = \frac{fv}{k}, \tag{II-10}$$

where $k = 2\pi/\lambda$. Here, of course, u and v are not spatial coordinates, but rather variables chosen to give this precise form of Eq. II-9.

C. INTERFERENCE AND COHERENCE

Interference is the name given to phenomena in which two or more waves participate and the effect of those waves is not the sum of the independent effects of each wave. The theory of interference is called " coherence theory." In the development of coherence theory, early misconceptions have been corrected, but many of these misconceptions are still widely held. Holography is based on interference, so a review of basic coherence is useful.

Early experimenters found that light derived from a single "point" source well filtered for monochromacity could be split into two beams which would exhibit interference effects. Such beams were called "coherent." Light from two sources or light which was not wave-length filtered exhibited no interference effects and was called "incoherent." We know now that these distinct classifications are misleading. There is a continuous variation from perfect coherence to perfect incoherence. The formalism for describing that variation is the partial coherence treatment of Born and Wolf.[1] Before we outline partial coherence theory, we must inquire into the origin of incoherence.

It is important to distinguish between incoherence arising from the wave source and incoherence arising from the interference experiment setup. The wave source will have random time variations which should be describable statistically. In addition, it may consist of an ensemble of smaller sources with some statistically describable correlation among them. Furthermore, the source will emit with a range of λ's. These effects comprise source limitations on the interference experiment. The existance of a spread in wavelengths leads to experimentally controllable coherence effects. If a certain amount of incoherence is tolerable, a more coherent source would permit relaxation of the experimental restrictions. For this reason the laser holograms reported by Leith and Upatnieks in 1962[2] aroused considerably more interest than the thermal-light holograms reported by Gabor in 1948.[3] Lasers allowed many tricks which would have been of very limited usefulness with thermal light.

Partial coherence theory is strictly a phenomenological method of describing interference effects. It does not reflect the origin of whatever incoherence is present. Remembering our definition of perfect coherence and perfect incoherence, we divide the observed interference pattern intensity (proportional to $|U|^2$ since physical detectors are all "square-law" detectors) into a coherent and an incoherent part. Thus the intensity is

$$I = |\gamma| I_{coh} + (1 - |\gamma|)I_{incoh}, \qquad (\text{II-12})$$

where I_{coh} and I_{incoh} are the intensities for these two limiting cases. Clearly, for all cases

$$O \le |\gamma| \le 1. \qquad (\text{II-12a})$$

The quantity $|\gamma|$, called the degree of coherence, depends on the observation time τ_0; on the difference, d, in path length between the two beams; and on the position, P, in the interference pattern. The depth of modulation, V, of an interference arising between two equally intense waves is related to the degree of coherence, $|\gamma|$, between the two waves by the equation

$$V = \frac{I_{max} - I_{min}}{I_{max} + I_{min}} = |\gamma|. \qquad (\text{II-13})$$

The degree of coherence is defined as the normalized absolute value of another quantity—the mutual coherence. For our purposes, however, $|\gamma|$ is the quantity of interest, and we will not pursue the subject of the mutual coherence itself.

We assume that the time variations in the wave at the point of measurement of the interference pattern obey certain statistical rules. We usually require that the interference pattern have a spatially nonuniform time average which is identical for all time-averaging intervals greater than

$$\tau_0 = \frac{2\pi}{\Delta\omega}. \qquad (\text{II-14})$$

The quantity τ_0 is the "coherence time." Such patterns are said to be "ergodic" (obeying some "rule" over all sufficiently long sampling intervals) and "stationary" (the "rule" being independent of the sampling interval's origin in time).

The finite bandwidth of all wave sources leads to restrictions on the interference experiments. The most severe experimental restriction arises from the necessity of matching the path lengths of the two beams to within a distance, ℓ, such that

$$\ell \ll L_c = c\tau_0 = \frac{2\pi c}{\Delta\omega}. \qquad (\text{II-15})$$

The quantity L_c is called the "coherence length." A path mismatch, l, will cause a phase mismatch, $(2\pi l \,\Delta\omega/c)$, between two rays with wavelengths ω and $\omega + \Delta\omega$. This mismatch causes the patterns for the various wavelengths to be misaligned. If $(2\pi l \,\Delta\omega/c) = (l/L_c)$ approaches 1, there will be an interference maximum for some ω at each point in space.

D. WAVE SOURCES

Potentially any type of wave can lead to an image with holography. This potential has been realized with many types of waves. The three basic types of waves which have been studied for holography are electromagnetic, acoustic, and particle waves.

Electromagnetic theory is based on Maxwell's equations. We can combine Maxwell's equations in such a way as to show that the electric and magnetic field vectors are transverse to the propagation direction. It is useful to assume a single, fixed polarization state so that the scalar wave equation (Eq. II-1) applies for either the electric or the magnetic field. The power of an electromagnetic wave is proportional to the square of either field. For this reason many authors use neither the electric nor the magnetic field, but a mathematical field, U (proportional to both the electric and the magnetic fields), which has the property that $|U|^2$ is the power of the radiation. The range in wavelengths of electromagnetic radiation stems from "too small to record wavelength size patterns with available media" to "too large to record on convenient records."

Polarization is an important consideration for electromagnetic holography. Three important results of polarization theory bear strongly on holography. First, electromagnetic radiation intensity, I, can always be described as having a polarized part, I_P, and an unpolarized (or randomly polarized) part, I_{UP}. The degree of polarization, μ, is defined in analogy to the degree of coherence, $|\gamma|$. Thus

$$I = (1 - \mu)I_{UP} + \mu I_P. \tag{II-16}$$

Second, the polarized waves can always be considered as consisting of waves of any arbitrary polarization state plus waves of the orthogonal polarization state. Third, only waves of identical polarization can produce an interference pattern. The third principle would virtually exclude all interference were it not for the second principle.

For sound, the quantity describable by the wave equation is the density. Clearly density is a scalar quantity, so no qualifications about polarization need to be added.

For particle waves, for example electron waves, the quantum mechanical wave function, ψ, is given by the equation

$$\nabla^2 \psi - \frac{1}{c^2} \frac{\partial^2 \psi}{\partial t^2} = m_0{}^2 c^2, \tag{II-17}$$

where m_0 is the rest mass of the particle. Equation II-17 is an inhomogeneous wave equation called the Schrödinger equation. For photons (electromagnetic particles) $m_0 = 0$, so the ordinary wave equation results. The means by which physical quantities are interpreted with the assistance of the wave function, ψ, is very complicated. Once again, though, the measurable quantities are more closely related to $|\psi|^2$ than to ψ.

For all waves, there arc two possible types of sources—spontaneous emission and stimulated emission. Spontaneous emission sources involve making some medium unstable with respect to emission of the particular type of wave and allowing the waves to be emitted spontaneously. Stimulated emission sources involve making some medium metastable with respect to emission of the particular type of wave and stimulation emission of those waves with an initial trigger wave (which may itself be spontaneously emitted). In general, the stimulated emission is essentially just an intensity-amplified version of the trigger wave; therefore stimulated emission sources can give much greater coherence than spontaneous emission sources. Examples of spontaneous emission sources are arc lamps, X-ray tubes, piezoelectric phonon sources, and electron guns. Examples of stimulated emission sources are lasers, masers, "phonon lasers," and the yet-to-be-developed "X-ray laser."

By far the most useful wave source for holography has been the laser. To date, only three types of lasers have been used extensively for holography. These are the ruby laser, the He–Ne laser, and the argon laser (all in the visible spectrum). Furthermore, not just any laser from these categories is useful. Usually the coherence is sufficient for three-dimensional holography only if the resonant cavity of the laser is so arranged that the only emission is in the TEM_{00} mode. That is the laser should be so designed that the only light emitted is propagated along the axis of the laser. The introduction of off-axis modes drastically reduces coherence length of the laser light. Even greater usefulness can be gaincd by the use of coherence-extending devices inserted into the resonant cavity.[4-7]

REFERENCES

1. M. Born and E. Wolf, *Principles of Optics*, Pergamon Press, London, 1959.
2. E. N. Leith and J. Upatnieks, *J. Opt. Soc. Am.*, **52**, 1123 (1962).
3. D. Gabor, *Nature*, **161**, 777 (1948).
4. L. Cirkovic, D. E. Evans, M. J. Forrest, and J. Katzenstein, *Appl. Opt.* **7**, 981 (1968).
5. H. P. Barber, *Appl. Opt.*, **7**, 559 (1968).
6. L. H. Lin and C. V. Lo Bianco, *Appl. Opt.*, **6**, 1255 (1967).
7. M. D. Domenico, Jr., *Appl. Phys. Letters*, **8**, 20 (1966).

III

Split-Beam Holography

A. THE ORIGIN AND AIMS OF SPLIT-BEAM HOLOGRAPHY

Historically, split-beam holography was introduced to separate the two images of a reconstructed hologram from each other and from the reconstructing beam (Section I-A). The split-beam technique is restricted to highly coherent sources because the path-matching is no longer almost automatic, as it was in the case of the Gabor hologram. This is the reason for the 14-year delay from Gabor's papers[1,2] to the Leith and Upatnieks papers[3,4]. The proper source was not available before the invention of lasers.

The basic idea of split-beam holography is to split the beam into two parts, only one of which is allowed to interact with the object. The waves from the object and the waves bypassing the object comprise the object and reference beams respectively. The two beams strike the plate at a nonzero relative angle, as shown in Figure III-1a. In the figure we choose the x-axis along the direction of the reference beam. A recording medium placed anywhere in the overlap region (shaded in Figure III-1a) will record a hologram. Reconstructing the hologram with a duplicate of the reference beam leads to the situation shown in Figure III-1b. Part of the reconstruction beam continuesundeflected. Part of the reconstructed beam forms a continuation of the object beam which produces a virtual image of the object. Another beam in a third direction also contains object information. This beam forms a distorted (pseudoscopic) real image.

In this chapter we will discuss the geometry of construction and reconstruction and distribution of energy among the three beams.

17

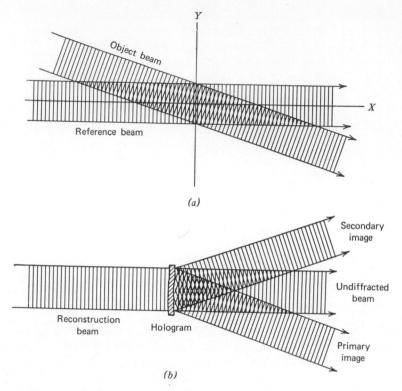

(a)

(b)

Figure III-1   The split-beam hologram method: *(a)* construction of the hologram *(b)* reconstruction of the object wavefront using the hologram from *(a)*.

B. THE INTERFERENCE PATTERN

We want to describe the interference pattern produced by a reference beam and an object beam. For the moment, we will assume that both beams are plane waves (Figure III-2*a*). We choose the *x*-axis as the bisector of the two beams. Let the reference field be

$$\exp(i\omega t)A_R \exp\left[ik(x\cos\theta - y\sin\theta)\right] \tag{III-1}$$

and the object field be

$$\exp(i\omega t)A_O \exp\left[ik(x\cos\theta + y\sin\theta)\right]. \tag{III-2}$$

The $\exp(i\omega t)$ terms will cancel out when we calculate intensities, so we simply omit them hereafter. The total field (in the region of overlap between the two beams) for perfect coherence is

$$A_T = A_R \exp\left[ik(x\cos\theta - y\sin\theta)\right]$$
$$+ A_O \exp\left[ik(x\cos\theta + y\sin\theta)\right]. \tag{III-3}$$

It is convenient to rewrite A_T in the form

$$A_T = A_O\left[1 + \left(\frac{A_R}{A_O}\right)\exp\left(-i2ky\sin\theta\right)\right]\exp\left[ik(x\cos\theta + y\sin\theta)\right]. \quad \text{(III-4)}$$

The intensity of the interference pattern is

$$|A_T|^2 = A_O{}^2\left[1 + \left(\frac{A_R}{A_O}\right)^2 + 2\frac{A_R}{A_O}\cos\left(2ky\sin\theta\right)\right]. \quad \text{(III-5)}$$

We recognize this as a dc term $A_O{}^2[1 + (A_R/A_O)^2]$ plus a cosine modulation term of magnitude $A_O{}^2\,2(A_R/A_O)$. The depth of modulation is

$$M = \frac{2(A_R/A_O)}{1 + (A_R/A_O)^2}. \quad \text{(III-6)}$$

The spacing in the y direction between maxima is

$$p = \frac{\lambda}{2\sin(\theta)}. \quad \text{(III-7)}$$

C. THE HOLOGRAM

For the moment we assume two physically unrealistic properties of the recording medium. First, we assume that the transmission, T, of the field vector through the recorded interference pattern is a linear function of $A_T{}^2$, that is, $T = a - bA_T{}^2$, where a and b are constants. This will be called "linear recording." Second, we assume that the recording medium is a flat two-dimensional surface. While there is no such thing as a two-dimensional medium, this approximation is often valid (Section V-D). A linear, two dimensional recording with the plane of the recording inclined at angle ϕ (Figure III-2b) with respect to the interference maxima (i.e., to the bisector of the reference and object beams) will produce a patterned structure with a cosine modulation of depth

$$M = \frac{2(A_R/A_O)}{1 + (A_R/A_O)^2} \quad \text{(III-8)}$$

and spacing between maxima of

$$S = \frac{\lambda}{2\sin\theta\sin\phi}. \quad \text{(III-9)}$$

This recorded interference pattern is a hologram.

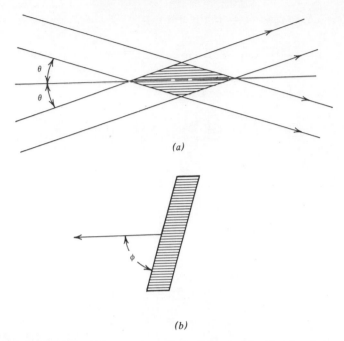

(a)

(b)

Figure III-2 Holography with two plane waves separated in direction by an angle 2θ: (*a*) the interference pattern and (*b*) the intersection of the interference pattern with a planar recording medium.

D. THE RECONSTRUCTION PROCESS

In order to reconstruct the object beam, we allow a duplicate of the reference beam to strike the hologram. We can calculate the effect of the hologram $T = a - b|A_T|^2$ on the reconstructing wavefront simply by multiplying the reconstructing wavefront by the hologram transmission, T. In this case the wavefront produced by a reconstructing wavefront identical to the reference wavefront is

$$A_R \exp\left[ik(x \cos \theta - y \sin \theta)\right][a - b|A_T|^2]. \qquad \text{(III-10)}$$

The first part of the expression is a multiplied by the incident wavefront and is of little interest. The second part is $-b$ multiplied by

$$A_R|A_T|^2 \exp ik(x \cos \theta - y \sin \theta).$$

It is convenient to set $a = 0$ and $-b = 1$ for calculation purposes and to rewrite Eq. III-5 in the form

$$|A_T|^2 = A_0{}^2 \left[1 + \left(\frac{A_R}{A_O}\right)^2 + \left(\frac{A_R}{A_O}\right) \exp\left(i2ky\sin\theta\right) \right.$$

$$\left. + \left(\frac{A_R}{A_O}\right) \exp\left(-i2ky\sin\theta\right) \right].$$ (III-11)

The reconstructed wavefront is then

$$A_R A_O{}^2 \left[1 + \left(\frac{A_R}{A_O}\right)^2 \right] \exp\left[ik(x\cos\theta - y\sin\theta)\right]$$

$$+ A_R A_O{}^2 \left(\frac{A_R}{A_O}\right)^2 \exp\left[ik(x\cos\theta + y\sin\theta)\right]$$

$$+ A_R A_O{}^2 \left(\frac{A_R}{A_O}\right)^2 \exp\left[ik(x\cos\theta - 3y\sin\theta)\right].$$ (III-12)

Let us now examine the three terms of Eq. III-12. The first term is a plane wave in the direction of the reconstructing beam. This is the undeviated beam. The second term is a plane wave in the direction of the original object beam. This is reconstruction of the object beam and is the primary (sometimes called "virtual") image beam. The third term is a plane wave deflected in the opposite direction from the primary image beam. This forms the conjugate (sometimes "real") image. In the case of a split-beam hologram of an actual object, the primary image is virtual and the conjugate image is real. For many other types of holograms both images are real. For the plane-wave hologram just described, the terms "virtual," "real," and "image" are difficult even to define. This accounts for our preference for the words "primary" and "secondary."

Physically, the hologram is a diffraction grating with a grating spacing $\lambda/2\sin\theta\sin\phi$. The reconstruction beam is split into an undiffracted beam and two first-order diffracted beams. The first-order beams are the primary and conjugate image beams.

E. GENERALIZED WAVEFRONTS

The properties just derived for plane wavefront holograms can be extended readily to generalized wavefronts by considering the generalized wavefront as consisting of a distribution of plane waves using the Fourier integral approach common in image analysis.[5] Thus a hologram formed by an object wavefront U_O and a reference wavefront U_R is governed by the analog of Eq. III-5

$$|U_T|^2 = |U_O + U_R|^2$$
$$= |U_O|^2 + |U_R|^2 + U_R U_O^* + U_R^* U_O. \tag{III-14}$$

If we have a linear recording medium, then

$$T = a - b|U_T|^2. \tag{III-15}$$

The term $|U_O|^2 + |U_R|^2$ governs the undeviated reconstruction beam. The term $U_R U_O^*$ governs the real image beam and the term $U_R^* U_O$ governs the virtual image beam.

The reconstruction of the hologram with wavefronts differing from U_R has been discussed by Sroke et al.[6,7] for the case of a point-source reference beam and an extended-source reconstructing beam and by Caulfield[8] for the case of an extended-source reference beam and a point-source reconstructing beam. In both cases much of the original information can be retrieved.

Experimentally, we find the image quality of a hologram to be very sensitive to its alignment with respect to the reference beam. A few minutes of arc rotation about either a horizontal or a vertical axis can produce noticeable changes in the clarity of images which are near the diffraction limits.

If the reconstruction beam differs from the reference beam* in wavelength, the image will change size and be slightly distorted in its details.[9]

If the point-source generating the reconstructing beam is not the same distance from the hologram as was the point-source for the reference beam, the resulting image will be magnified or demagnified.[10]

These results can be summarized very simply. If the reconstructing beam is not an exact duplicate of the reference beam, there will not be the most exact possible correspondence between the image and the object.

REFERENCES

1. D. Gabor, *Nature*, **161**, 777 (1948).
2. D. Gabor, *Proc. Roy. Soc.*, (*London*) **A197**, 454 (1949).
3. E. N. Leith and J. Upatnieks, *J. Opt. Soc. Amer.*, **52**, 1123 (1962).
4. E. N. Leith and J. Upatnieks, *J. Opt. Soc. Amer.*, **53**, 1377 (1963).
5. G. Wade, *IEEE Trans. Sonics Ultrasonics* **15**, 51 (1968).
6. G. W. Stroke, R. Restrick, A. Funkhouser, and D. Brumm, *Phys. Letters*, **18**, 274 (1965).
7. G. W. Stroke, R. Restrick, A. Funkhouser, and D. Brumm, *Appl. Phys. Letters*, **7**, 178 (1965).
8. H. J. Caulfield, *Phys. Letters*, **27A**, 319 (1968).
9. R. W. Meir, *J. Opt. Soc. Amer.*, **55**, 987 (1965).
10. J. B. Develis and G. O. Reynolds, *Theory and Applications of Holography*, Addison-Wesley, Reading, Mass., 1967, Chapters 4 and 8.

* In the limit of plane-wave reference and reconstruction beams, these effects go to zero.

IV

Single-Beam Holography

A. THE PURPOSES OF SINGLE-BEAM HOLOGRAPHY

Single-beam holography is what you use if, for some reason, you do not want to split part of the beam away before it hits the object or if you just do not have a beam splitter. This was the problem facing Gabor when he sought to make a hologram with electron waves. In one of his early papers[1] Gabor states:

> In light optics a coherent background can be produced in many ways, but electron optics does not possess effective beam-splitting devices; thus the only expedient was in using the illuminating beam itself as the coherent background.

Thus single-beam holography is the art of deriving the reference beam from the object beam.

Other than convenience, when beam splitters are inconvenient or unavailable, another advantage usually stems from single-beam holography. Since the reference and object beam are identical before they strike the object, it is usually possible to achieve path matching (Section II-B) for objects at arbitrary distances. Thus if objects at unknown distances much greater than the coherence distance of the source are to be used, single-beam holography is very helpful. For thermal sources the coherence distance is usually much less than a millimeter, so single-beam holography is necessary, for example, Gabor's holograms with thermal light.[1-3] For lasers the coherence distance can be quite long, but not long enough to allow split-beam holography of outdoor objects, so once again single-beam holography is useful.[4,5]

B. THE GABOR HOLOGRAM

The Gabor hologram[1-3] requires an object which "knocks a hole" in a larger coherent illuminating beam. For example, the object might be a small solid object in a larger beam. Most of the waves would continue undisturbed, but there would be an object-shaped "hole" where the object blocked part of the incident wavefront. Again the object might be a transparency. Most of the

23

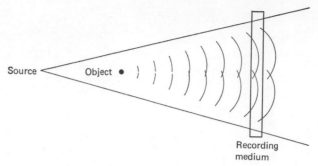

Figure IV-1   A typical Gabor hologram construction setup.

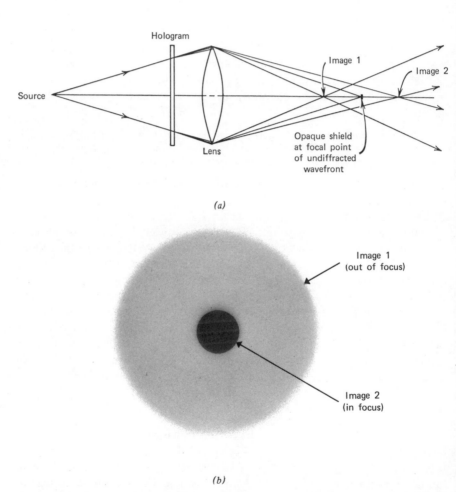

(a)

(b)

Figure IV-2   A typical Gabor hologram reconstruction setup (*a*) and the resulting image (*b*).

incident wavefront would be unperturbed, but some parts would be blocked by the opaque regions of the transparency.

The Gabor construction and reconstruction processes are shown in Figures IV-1 and IV-2, respectively. As noted in Chapter I, the two images are on axis and are obscured by the reconstructing beam. A small disc can be used to block the reconstructing reference beam at its focal point as shown in Figure IV-2a. Then the image (Figure IV-2b) contains an in-focus and an out-of-focus part.

If the object is diffusing, the images can be viewed from off axis as shown by Stroke et al.[6]

C. VARIATIONS ON GABOR'S METHOD

There have been numerous variations on Gabor's method—some of them by Gabor himself.[7,8] Of these the one of greatest interest, at the time of this writing, is the laser holographic particle sizing camera of Thompson et al.[9-11] The particle sizing camera will be discussed in Chapter XV. Historically, this was probably the first significant application of holography.

D. BUILT-IN REFERENCE BEAMS

It is possible to imagine objects which will carry with them mirrors to reflect part of the incident wave back toward the recording medium. In this case the reference beam is built-in. Similarly a bright point on the object can serve as a point reference beam source, thus building in the reference beam. The distinguishing mark of the built-in reference beam method is that some peculiarity of the object must be assumed, for example, the object must carry a mirror or be illuminated in a highly nonuniform way.[12,13]

A significant variation of these methods is "holography without separated reference beam," as described by Stroke et al.[14,15] They point out that the wavefront O at the object can be broken into parts A and B, where

$$O = A + B.$$

In the far field (see Chapter VI for discussion of "far field" and "near field"), the fields overlap. The far fields are U_O, U_A, and U_B for the total object, the A part, and the B part, respectively. An ordinary diffraction pattern for the object is a recording of

$$|U_O|^2 = |U_A + U_B|^2$$
$$= |U_A|^2 + |U_B|^2 + U_A U_B{}^* + U_A{}^* U_B. \qquad \text{(IV-1)}$$

Clearly, the diffraction pattern is a hologram! If U_A is known or can be approximated $|U_A|$, it can be used to reconstruct U_B. Thus the phase is not lost even without a separated reference beam. There is some hope that this technique may be useful in X-ray holography.[16]

E. LOCAL REFERENCE-BEAM HOLOGRAPHY

As we look back on the various forms of single-beam holography, we note that none of them offers the convenience of an ordinary camera. With an ordinary camera and a distant object there is only one requirement for image formation: the amount of light reaching the film during the time required to "stop" the object motion must expose the film. There is no need to know how far away the object is. There is no need to illuminate the object non-uniformly. There is no need for mirrors or other extraneous devices near the object. All we need is a short enough burst of light with enough energy in it to expose the film.

Caulfield et al[4,5] have recently devised a method called "local reference-beam (LRB) holography," which is as simple as ordinary photography. All that is required is that sufficient energy reach the recording medium during the time necessary to stop the object motion.

The basic idea of LRB holography is to use the waves from the object alone. Part of the wave is split off near the recording medium and condensed to serve as a reference beam. Thus the object beam generates its own reference beam. A typical LRB camera is shown in Figure IV-3. Part of the in-

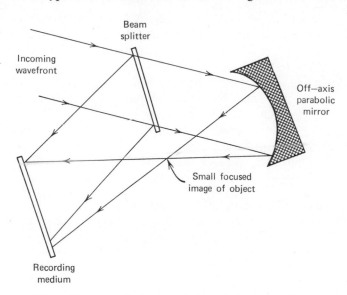

Figure IV-3 A typical reference beam holographic camera.

coming wavefront is focused by the curved mirror to form a very small image of the object. The small image serves as a pointlike source for the reference beam. Reconstruction is accomplished by illuminating the hologram with waves from a point source. Because the reconstructing wavefront is not a duplicate of the reference beam, distortions will result.

It is possible to describe the image formation with an extended-point-source reference beam by a very straightforward analysis. The following quantities are important:

d_1, the distance from the object (or aerial image of the object if intervening optics is to be used) to the recording medium;

d_2, the distance from the "point-like" source to the recording medium;

w_1, the wavefront reaching the recording medium from the object (or aerial image of the object);

w_1, the reconstructed object wavefront at the hologram;

w_2, the wavefront at the recording medium from the pointlike reference source;

w_2', the wavefront at the hologram from the reconstructing source;

Δx, the linear size of the extended reference-beam source;

Δy, the linear size of deviations of w_2 from perfect sphericity;

$\Delta \xi$, the smallest resolvable image detail;

λ, the constructing wavelength;

λ', the reconstructing wavelength.

We will approximate diffraction effects by assuming that an object of size Δu a distance d away from the image plane will produce a diffraction pattern at wavelength λ with a central lobe of width Δv given by

$$\Delta v \approx \frac{\lambda d}{\Delta u}. \tag{IV-2}$$

This holds for rectangular apertures and is approximately true for any aperture.

Thus, w_2 deviates from a spherical wave by

$$\Delta y \approx \frac{\lambda d_2}{\Delta x}. \tag{IV-3}$$

Reconstruction with a point source at the location of the extended reference-beam source produces a perfectly spherical w_2'. The image-producing term in the hologram transmission is

$$w_2^* w_1.$$

Reconstruction with w_2' leads to an object wavefront

$$w_1' = w_2' w_2^* w_1.$$

The term $w_2'w_2^*$ differs from a delta function by variations of size Δy. Thus, w_1' and w_1 differ over areas of size Δy. Therefore

$$\Delta\xi = \frac{\lambda' d_1}{\Delta y}$$

$$= \left(\frac{\lambda' d_1}{\lambda d_2}\right)\Delta x. \qquad\qquad \text{(IV-4)}$$

The information content of the image depends on the image size, ξ, divided by the resolution size, $\Delta\xi$. Let us call this ratio I.

$$I = \frac{\xi}{\Delta\xi}$$

The object size is assumed to be x, so we let

$$m = \frac{\xi}{x}$$

where m is the magnification. In general,

$$m = M\left(\frac{\lambda'}{\lambda}\right),$$

where M is geometrical magnification due to any lenses which may have been used. Thus,

$$I = \frac{\xi}{\Delta\xi} = (Mx)\frac{(\lambda'/\lambda)(\lambda/\lambda')(d_2/d)}{\Delta x}$$

$$= M\left(\frac{d_2}{d_1}\right)\left(\frac{x}{\Delta x}\right)$$

$$= M\left(\frac{d_2}{d_1}\right)I_0 \qquad\qquad \text{(IV-5)}$$

where $I_0 = x/\Delta x$. Equation IV-5 is the key expression. Note that any image magnification due to λ'/λ is empty in the sense that the resolution size also is magnified by λ'/λ. Figure IV-4 shows the reconstructed real image for the case $M = 1$, $\lambda'/\lambda = 1$, $d_1/d_2 = 3$, and $\Delta x \approx 0.1$ cm. According to Eq. IV-4, this sould lead to $\Delta\xi \approx 3$ mm. The trade name on the tape dispenser is written in letters slightly greater than 1 mm in stroke. Clearly the real value of $\Delta\xi$ is somewhere between 1 and 2 mm, in close agreement with this approximate theory.

The next major step in LRB holography was taken by Brandt.[17] He used a holographic lens to focus an image of the object onto the photographic plate (see Figure IV-5). The undeviated object beam interfered with the focused

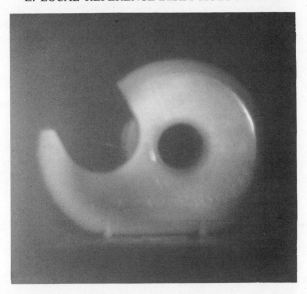

Figure IV-4 A photograph of the real image reconstructed from a local reference beam hologram. From Caulfield, *Phys. Lett.*, **27A**, 319 (1968).

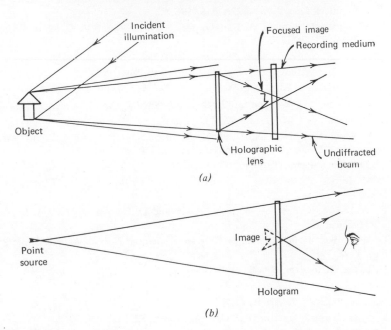

Figure IV-5 Brandt's modified version of local reference beam holography showing the construction (*a*) and reconstruction methods (*b*).

beam to produce a LRB hologram. This had the effect of setting $d_1 \approx 0$ in Eq. IV-4 to achieve high resolution. Thus, the known insensitivity to reference source size of the image plane hologram[18] makes LRB holography capable of high resolution.

Note that Brandt's hologram camera is very simple. This should be very helpful in producing diffraction-limited images and should make holography a technique useful in any laboratory with very little equipment.

Interestingly enough, a camera much like Brandt's was proposed and used for incoherent holography by Young[19] in 1963 (although Young was apparently unaware of its relationship to Gabor's work in 1948[1] or to that of Leith and Upatnieks[20] which had just been published). Holography is returning to its origins.

REFERENCES

1. D. Gabor, *Proc. Roy. Soc.* (*London*), **A197**, 454 (1949).
2. D. Gabor, *Nature*, **161**, 777 (1948).
3. D. Gabor, *Proc. Phys. Soc.*, (*London*), **B64**, 449 (1951).
4. H. J. Caulfield, J. L. Harris, H. W. Hemstreet, Jr., and J. G. Cobb, *Proc. IEEE* (*London*), **55**, 1758 (1967).
5. H. J. Caulfield, *Phys. Letters*, **27A**, 319 (1968).
6. G. W. Stroke, D. Brumm, A. Funkhouser, A. Labeyrie, and R. C. Restrick, *Brit. J. Appl. Phys.*, **17**, 497 (1967).
7. D. Gabor, *Proc. Inter. Congr. Electron Microscopy*, 1*st* **19**, 129 (1953).
8. D. Gabor and W. P. Goss, *J. Opt. Soc. Amer.*, **56**, 849 (1966).
9. B. J. Thompson, J. Ward, and W. Zinky, *J. Opt. Soc. Amer.*, **55**, 1566A (1965).
10. B. J. Thompson, G. B. Parrent, B. Justh, and J. Ward, *J. Appl. Meteorol.*, **5**, 343 (1966).
11. B. J. Thompson, J. Ward, and W. Zinky, *Appl. Opt.*, **6**, 519 (1967).
12. J. W. Goodman, NAS—NRC report, *Reconstruction of Atmospherically Degraded Images*, Vol. 1, Woods Hole Summer Study, July 1966, p. 61.
13. W. D. Montgomery, *Appl. Opt.*, **7**, 83 (1968).
14. G. W. Stroke, R. Restrick, A. Funkhouser, and D. Brumm, *Phys. Letters*, **18**, 274 (1965).
15. G. W. Stroke, R. Restrick, A. Funkhouser, and D. Brumm, *Appl. Phys. Letters*, **7**, 178 (1965).
16. G. W. Stroke, *An Introduction to Coherent Optics and Holography*, Academic Press, New York, 1966.
17. G. B. Brandt, *J. Appl. Opt.* **8**, 1421 (1969).
18. L. Rosen, *Appl. Phys. Letters*, **9**, 337 (1966).
19. N. O. Young, *Sky Telescope*, **25**, 8 (1963).
20. E. N. Leith and J. Upatnieks, *J. Opt. Soc. Amer.*, **52**, 1123 (1962).

V

Physical Recording Media and Reconstruction Efficiencies of Holograms

A. INTRODUCTION

A replica of the interference pattern produced by the wave scattered from the object of interest and the coherent reference wave must exist as detectable changes in some physical medium to constitute a hologram. In principle, the spatial variance of light intensity of the interference patterns can be recorded on a medium through any change of its physical properties, provided those properties can interact with the reconstruction beam so that an image can be formed. In most cases the interference pattern is recorded on the medium in a form of either an optical density pattern or a phase shift pattern. In the former case, the material is absorptive and the hologram recorded is a density grating which diffracts light by modulating the intensity of the reconstruction beam. In the latter case, the material is nonabsorptive or "phase-only," and the hologram has a uniform optical density. The phase variation of the recording media diffracts light just like a phase grating. For phase-only materials, the required phase shift of the reconstruction wave is provided through the local variations in the thickness and/or the refractive index of the recording medium. In many cases both amplitude and phase variations are present in the same hologram.

Among the various recording media that have been used for holographic recording are the following:

1. Photographic emulsions. Basically absorptive media. However, developed photographic emulsions can be bleached in a manner such that a phase-only image will remain. Phase holograms have been produced in this way using various photographic emulsions.

2. Photochromic materials. These are erasable materials that can be reused over and over. The materials contain absorptive color centers, and most

31

of them are sensitive only in the shorter wavelength region of the visible spectrum and in the ultraviolet. The resolution capability of these material is often very high and the scattering noise is very low. However, the sensitivity is also very low, so high power lasers or long exposure time is necessary.

3. Photopolymer materials. These include photosensitive resist and other organic materials which have the property of changing from monomer to polymer due to exposure to light. After exposure, the materials show either differential index of refraction or differential solubility to a solvent. The recorded holograms on these materials are therefore phase-only. Good quality holograms have been recorded on some green and blue sensitive resist, such as Kodak Ortho Resist. Also, with certain materials, holograms can be recorded without having chemical development and fixing.[1] Thus, the process has the advantage of achieving nearly real-time reconstruction of the recorded image.

4. Thermoplastic materials. Holograms have been recorded by means of thermoplastic xerography.[2] The material responds to electrostatic force and forms a phase image from the charge-induced deformation. The reported sensitivity is quite high, and the scattering noise is very low.

5. Dichromate sensitized gelatin. These are materials which are believed to be sensitized by a process of reducing the dichromate ion Cr^{6+} to Cr^{3+} by exposure to light to form a complex which hardens the gelatin A phase image is formed by dissolving away the unhardened gelatin. High resolution capability and noiseless recording have been reported.[3,4] The material is very easy to prepare and handle. For example, a gelatin plate can be prepared by fixing an unexposed Kodak 649F plate. Then it can be sensitized by soaking the plate in ammonium dichromate solution. The sensitivity of the dichromate-sensitized gelatin plate lies in the blue region of the visible spectrum. Holograms of over 95% efficiency and of excellent image quality can be achieved with this material.

6. Electrooptic crystals. The first crystal used for hologram storage was lithium niobate. A suitably intense beam from an argon laser produces refractive index changes in the crystal. By heating the crystal to 170°C, the hologram can be erased so that the crystal can be reused. This material gives high efficiency and low noise.

Table I summarizes the properties of recording media that have been used in holography.

On some occasions the interference pattern is recorded as a single layer of spatial grating, and the resultant hologram is referred to as a "single-layer

TABLE I   PROPERTIES OF HOLOGRAPHIC RECORDING MEDIA

Recording medium	Typical resolution, lines/mm	Typical exposure required, ergs/cm^2	Absorptive (A) or phase-only (P)
Photographic emulsions	Over 2000	$20-10^3$	A and P
Photochromic materials	Over 2000	10^5-10^7	A
Photopolymer materials	Over 1000	10^4-10^6	P
Thermal-plastic materials	1000	$10-100$	P
Dichromate-sensitized gelatin	Over 2000	10^5	P
Electrooptic crystals	Over 4000	10^9	P

hologram." In other cases the recording medium is many wavelengths thick. The three-dimensional spatial interference pattern is recorded as a three-dimensional grating. The resultant hologram is a "volume hologram." A volume hologram can be either of a transmitting type or a reflective type. In both cases Bragg diffraction plays an important role in the reconstruction process.

In the following sections, we describe the behavior of a thin photographic emulsion as a holographic recording medium and then discuss the theoretical reconstruction efficiencies of various hologram types.

B. PHOTOGRAPHIC EMULSIONS AS HOLOGRAM RECORDING MEDIA

Photographic emulsion is an absorptive media according to the classification given above. When the material is exposed to the light, developable centers are formed on the silver halide grains inside the emulsion. In the development process, silver grains are formed from these centers. The portion of the emulsion which receives more exposure will have more silver grains developed. After completion of the fixing process, which removes the unexposed sensitive silver halide in the emulsion, a negative black and white image remains. The development process can be considered as *a posteriori* amplification of the inherent photosensitivity. The larger the grains, the greater the amplification and (unfortunately for holography) the lower the resolution. Holography usually requires high resolution and, hence, relatively low amplification. Nevertheless, this low amplification suffices to make photographic media orders of magnitude more sensitive to light than nonphotographic (nonamplified) media of comparable resolution. The combination of this sensitivity with widespread availability and ease of handling make photographic media by far the most popular holographic recording media. Before we describe the response of photographic emulsions as hologram recording media, it is worthwhile to review briefly the ordinary photographic sensitometry of the material.

1. The H-D Curve

The exposure characteristic of photographic emulsion is given by the H-D (Hurter-Driffield) curve of the optical density versus the common logarithm of the exposure. The optical density, D, is the common logarithm of the opacity, O; and the opacity is defined as the inverse of the light intensity transmission, that is (in terms of the transmission, T, for the field vector).

$$O = \frac{1}{T^2}$$

$$D = \log_{10}(O) = \log_{10}\left(\frac{1}{T^2}\right) = -2\log_{10} T. \qquad \text{(V-1)}$$

A typical H-D curve is shown in Figure V-1. The linear portion of the H-D

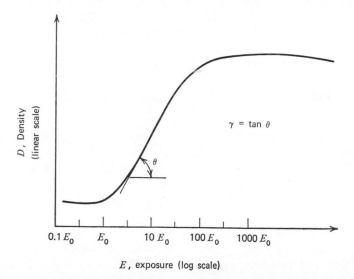

Figure V-1   A typical Hurter-Driffield Curve for a photographic recording medium.

curve is often referred to in photography as the range for correct tone re-production. The slope of this linear portion is the gamma (γ) of the developed emulsion. For a given material, the value of γ can be controlled by changing the type of developer and the time of development.

2. Single-Layer Hologram Recorded on Photographic Emulsion

Let us assume that we record a single layer hologram on a photographic emulsion. In chapter III we illustrated the theory of wavefront reconstruction by assuming the recording process of the hologram is linear. In reality, it is noted from the H-D curve and the relationships between D and T that the holographic recording process on photographic emulsion is far from linear. In formulating a theory of wavefront reconstruction from a real photographic emulsion, some authors have assumed that the exposure of the hologram is in the linear region of the H-D curve. This is seldom the case for a correctly exposed hologram. For example, Friesem et al.[5] have determined the condition for an optimum reconstruction of a hologram. They find that the maximum signal to noise ratio of the reconstructed image happens in a hologram with an average field transmission $T = 0.5$, which corresponds to a density of approximately 0.6. For most photographic emulsions used for hologram recording, such a density is well in the toe region of the H-D curve.

A more sophisticated theory can be formulated by borrowing some concepts from communication theory and using the curve of field transmission, T, versus the exposure, E. This curve is the transfer characteristic of the hologram medium, and the exposure received by the hologram is the signal recorded. Let us use a simple example to illustrate the theory. A hologram with a plane reference wave and a plane object wave has an exposure which is

$$E = [I_R + I_O + 2\sqrt{I_R I_O} \cos (Ky)]t \qquad (V-2)$$

where I_R and I_O are the intensities of the reference and object waves, respectively, K is a constant determined by the incident angles of both waves, and t is the exposure time. Equation V-2 can be obtained from Eq. III-5 by letting

$$I_R = |A_R|^2$$
$$I_O = |A_O|^2$$
$$K = 2k \sin \theta.$$

The depth of modulation of such a signal is

$$M = \frac{2\sqrt{I_R I_O}}{(I_R + I_O)}. \qquad (V-3)$$

The dc term in the exposure given by $E_O = (I_R + I_O)t$ is a bias and also the average exposure of the hologram. The ac component $2\sqrt{I_R I_O}\, t \cos (Ky)$ (as is shown in Chapter III) forms the reconstructed wavefronts. Therefore the depth of modulation has the physical meaning of the percentage of image-forming (or useful) information recorded on a hologram.

A typical transfer characteristic (T vs E curve) of a hologram recorded on a Kodak 649F plate is shown in Figure V-2. When the exposure is zero, the field transmission of the emulsion is 100%. As the exposure increases, the transmission decreases gradually until the exposure is so high that the emulsion appears totally black and the field transmission, T, drops to zero.

It is also shown in Figure V-2 that the dc component of the exposure E_0

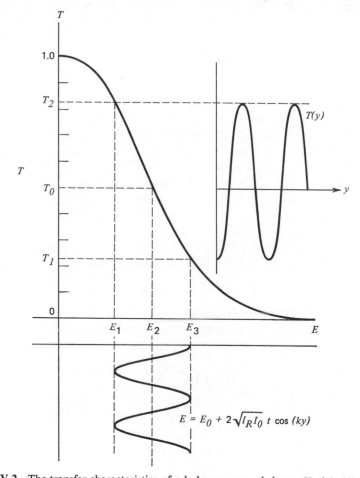

Figure V-2 The transfer characteristics of a hologram recorded on a Kodak 649 F plate.

will cause a constant field transmission level T_0, and the ac component $2\sqrt{I_R I_0}\, t \cos (Ky)$ causes a corresponding swing of T around T_0. However, due to the nonlinearity of the transfer characteristic, the swing of T is not sinusoidal. We expand $T(E)$ into a Taylor series around E_0, that is,

$$T(E) = T_0 + \left.\frac{dT}{dE}\right|_{E_0} (E - E_0) + \frac{1}{2!} \left.\frac{d^2T}{dE^2}\right|_{E_0} (E - E_0)^2$$

$$+ \text{ higher order terms,} \tag{V-4}$$

and let

$$E - E_0 = [2\sqrt{I_R I_O} \cos (Ky)]t$$

and

$$\beta = \left.\frac{dT}{dE}\right|_{E_0}.$$

We then have

$$T(y) = T_0 + 2\beta t \sqrt{I_R I_O} \cos (Ky)$$

$$+ 2 \left.\frac{d^2T}{dE^2}\right|_{E_0} t^2 I_R I_O \cos^2 (Ky)$$

$$+ \text{ higher order terms.} \tag{V-5}$$

It is obvious that in the reconstruction process, the first term is a dc transmission, which only attenuates the reconstruction beam. The second term is a sinusoidal grating, which diffracts light and forms the reconstructed wavefronts. The next term can| be modified by a trigonometric identity to read

$$\left.\frac{d^2T}{dE^2}\right|_{E_0} t^2 I_R I_O [1 + \cos (2Ky)]. \tag{V-6}$$

Equation V-6 contains a dc term and a sinusoidal grating of double spatial frequency which will cause diffractions at larger angles usually known as the second-order diffractions. The next higher order term contains $\cos^3 (Ky)$. By using more trigonometric manipulations, we have

$$\cos^3 (Ky) = \frac{[\cos (3Ky) + 3 \cos (Ky)]}{4}. \tag{V-7}$$

We observe that Eq. V-7 has a term $3 \cos (Ky)/4$ that will contribute some to the first-order reconstructed wavefronts, and a term $\cos (3Ky)/4$ that contributes to the third-order diffractions.

In fact, in this special case of a hologram of two plane waves, all the higher order terms containing the odd powers of $\cos (Ky)$ have some contribution to the first-order reconstructed wavefronts. A general hologram of arbitrary wavefronts will not have contributions from higher order terms to the reconstructed wavefronts. However, the higher order diffraction still exists.

If we neglect all the higher order terms in Eq. V-5 and only keep the first two, then we have

$$T(y) = T_0 + 2\beta t\sqrt{I_R I_O} \cos(Ky).$$

From the definition of the depth of modulation, we can write

$$T(y) = T_0 + \beta E_0 M \cos(Ky)$$

$$= T_0 + \frac{\beta E_0 M(e^{iky} + e^{-iky})}{2}. \tag{V-8}$$

If the reconstruction beam of amplitude U_R' is used to illuminate the hologram, then each reconstructed wavefront has an amplitude which is

$$\frac{\beta E_0 M U_R'}{2}.$$

The intensity of a reconstructed wavefront (either the virtual or real image) divided by the total intensity of the reconstruction beam is the reconstruction efficiency of a hologram. In this example we have

$$\eta = \text{reconstruction efficiency} = \frac{\beta^2 E_0^2 M^2}{4} \tag{V-9}$$

In Eq. V-9, β is the slope of the transfer characteristic at E_0 and thus is a function of E_0. Therefore, at a constant dc exposure E_0, the reconstruction efficiency is proportional to the square of the depth of modulation.

The reconstruction efficiency can be written as a function of the familiar reference to object intensity ratio. If we call $I_R/I_O = R$, then

$$M^2 = \frac{4R}{(1 + R)^2}$$

and at a fixed E_0, we have

$$\eta \propto \frac{R}{(1 + R)^2}. \tag{V-10}$$

Figure V3 shows the variation of the reconstruction efficiency versus the reference-to-object intensity ratio, R. It is noted from the curve that the use of a low ratio results in a high reconstruction efficiency. In the meantime, it increases the swing of T shown in Figure 2 to the highly nonlinear region such that the higher order diffractions can be quite strong and may severely degradate the reconstructed image. Therefore, a ratio of $R \approx 3$ is usually chosen as the best compromise between efficiency and image quality.

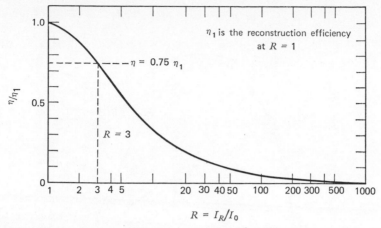

$$R = I_R/I_0$$

Figure V-3   The reconstruction efficiency, η, for fixed β and E_0 varies with the reference-to-object intensity ratio, R.

When the depth of modulation, M, is fixed, the reconstruction efficiency is proportional to $(\beta E_0)^2$. Friesem et al.[4] have found that with Kodak $649F$ plate, this product approaches maximum when the dc field transmission, T_0, is about 0.5. Figure V-4 shows the variation of the reconstruction efficiency as a function of the dc field transmission.

For a hologram which records the general wavefronts, the exposure is given by Eq. III-14. We can extend the theory by defining the depth of modulation as the useful information recorded divided by the average exposure of the hologram, that is

$$M = \frac{2|U_R||U_O|}{(|U_R|^2 + |U_O|^2)}. \tag{V-11}$$

However, in this case the depth of modulation is not a constant over the hologram plane. The hologram can be divided into many small holograms of constant depth of modulation. Each small hologram diffracts light during the reconstruction. The total diffraction is the total of the contributions of these small holograms. The reconstruction efficiency is then proportional to the mean square value of M, that is,

$$\eta \propto \langle M^2 \rangle, \tag{V-12}$$

In holograms which record transparencies and in many Fourier transform holograms, the distribution of M over the hologram plate is far from uniform. In order to have linear recording over the entire plate, the proper reference to object intensity ratio, R, is chosen for the portion of the hologram where the depth of modulation is maximum. In this case the rest of the hologram, where the depth of modulation is much smaller, will have a high reference-to-object intensity ratio such that they do not contribute much light to the

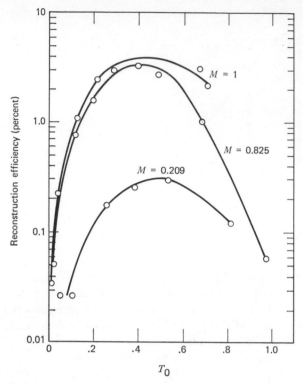

Figure V-4   The variation of reconstruction efficiency, η, with the DC field transmission, T_0, for Kodak 649 F plates. After Friesem et al.[4].

reconstructed image. The total efficiency of the hologram is low. In other words, such a hologram does not have a very efficient use of the dynamic range of the recording medium.

When a diffuser is inserted in to the object beam or a diffused object is recorded, the distribution of the depth of modulation over the hologram plate is much more uniform. This generally results in a hologram with higher reconstruction efficiency.

The dependence of η on the dc exposure E_0 is the same as in the previous case. The amount of exposure needed to obtain the optimum dc field transmission level on a given photographic emulsion depends on the developing process and must be determined experimentally.

C. WAVEFRONT RECONSTRUCTION FROM A SINGLE-LAYER, PHASE-ONLY HOLOGRAM

The theory of wavefront reconstruction from a single-layer, phase-only hologram is slightly different from those described in the previous section

As before, we will use the case of a plane reference wave and a plane object wave to illustrate the theory. The exposure given in Eq. V-2 is recorded as a phase change of the medium.

If the recording material has a linear response, that is, the phase change of the medium caused by the exposure of light is proportional to the amount of exposure received, then for the sinusoidally varying exposure given in Eq. V-2, the resultant phase shift on the hologram is also sunusoidal. The field transmission function of the hologram is therefore

$$T(y) = e^{-i\phi(y)}$$

and

$$\phi(y) = \phi_0 + \phi_1 \cos (Ky). \tag{V-13}$$

If we use the well known Bessel expansions of $\sin [\phi_1 \cos (Ky)]$ and $\cos [\phi_1 \cos (Ky)]$ functions, we can write the field transmission function of the hologram as:

$$T(y) = e^{-i\phi_0} e^{-i\phi_1 \cos (Ky)}$$

$$= e^{-i\phi_0} \{\cos [\cos (\phi_1 Ky)] - i \sin [\cos (\phi_1 Ky)]\}$$

$$= e^{-i\phi_0} \left[J_0(\phi_1) + 2 \sum_{n=1}^{\infty} \frac{J_n(\phi_1) \cos (nKy)}{i^n} \right]. \tag{V-14}$$

When the hologram is reconstructed, as it is shown in the previous section, the dc term $J_0(\phi_1)$ is the undiffracted wave, and the term

$$\frac{2}{i} J_1(\phi_1) \cos (Ky) \tag{V-15}$$

is a sinusoidal grating which diffracts light and forms the reconstructed wavefronts. The rest of the higher order terms represent sinusoidal gratings of double, triple, and higher multiple spatial frequencies. Their effect is to cause higher order diffractions.

Let us find out the efficiency of a single-layer phase-only hologram. From Eq. V-14, if we have a reconstruction beam of unit amplitude, the reconstructed wavefronts are

$$\frac{2}{i} J_1(\phi_1) \cos (Ky) = \frac{1}{i} J_1(\phi_1)(e^{iKy} + e^{-iKy}). \tag{V-16}$$

Hence, each reconstructed wavefront has an intensity given by $J_1^2(\phi_1)$. Therefore, the reconstruction efficiency of the hologram is

$$\eta = J_1^2(\phi_1). \tag{V-17}$$

Figure V-5 shows the reconstruction efficiency as a function of ϕ_1. It is noted that the reconstruction efficiency increases with respect to ϕ_1, reaches a

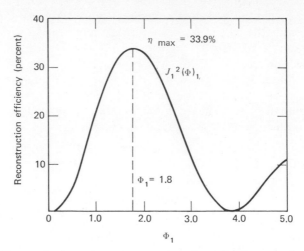

Figure V-5 The graph shows how the reconstruction efficiency, η, of a single-layer, phase-only hologram varies with the magnitude, ϕ_1, of the phase variation.

maximum of 33.9 % at $\phi_1 = 1.8$, then drops back again for further increasing of ϕ_1.

In the analysis given above, we have assumed a recording medium which has a linear relation between $\phi(y)$ and $E(y)$. Under this assumption, ϕ_1 is proportional to the depth of modulation, M, of the recorded signal, that is,

$$\phi_1 = \alpha M, \tag{V-18}$$

where α is a proportionality constant. Therefore,

$$\eta = J_1^2(\alpha M). \tag{V-19}$$

Equation V-19 gives the reconstruction efficiency as a function of the depth of modulation. When the depth of modulation is fixed, the efficiency depends on the proportionality constant. This parameter, which corresponds to the parameter $(\beta E)^2$ in Eq. V-9, depends on the recording material and the ways it is treated.

If we compare the properties of a single-layer phase-only hologram to those of a single-layer absorptive hologram, it is noted that the phase-only hologram has more intense higher order diffractions. When the recording medium has a linear relation between the resultant phase change and the exposure received, the hologram is a sinusoidal phase grating like those shown in Eq. V-13. However, a sinusoidal phase grating results in strong higher order diffractions through the Fourier expansion given by Eq. V-14. This is quite different from a sinusoidal absorptive grating whose characteristics were discussed in the previous section.

D. MAXIMUM RECONSTRUCTION EFFICIENCIES OF HOLOGRAMS

We have just discussed the properties of the single-layer holograms and their reconstruction efficiencies as functions of the depth of modulation and the dc exposure level. In this section we are going to discuss the ultimate reconstruction efficiencies of various types of holograms. As was mentioned previously, a hologram which records general wavefronts can be divided into many small holograms, each of which records a sinusoidal interference pattern of constant depth of modulation. The reconstruction efficiency of such a hologram is given by the average value of the reconstruction efficiencies of individual small holograms. In this sense, a hologram of maximum reconstruction efficiency requires that each small hologram also have the maximum reconstruction efficiency and, thus, an equal depth of modulation. This can occur only when two plane waves interfere. Therefore, we shall describe the theory of a hologram which records a sinusoidal exposure of constant depth of modulation. The reconstruction efficiency can be maximum only if the exposure is everywhere uniform, that is, only for holograms formed by two plane waves. Slight deviations lead to slight degradations in efficiency. Completely diffuse object wavefronts lead to about 50% degradations. The reconstruction efficiency is defined as the ratio between the amount of diffracted light which contributes to the reconstructed image and the intensity of the reconstruction wave. Since the nonlinearity due to the recording media generally only causes the higher order diffractions which contribute nothing to the reconstruced image, the recording process is therefore assumed to be linear throughout the discussion.

We shall discuss the cases of the single-layer holograms and the volume holograms separately. In the former case, the information is recorded as a single layer on the medium such that a two-dimensional theory is adequate to describe the behavior of the hologram. In the case of volume holograms, the recording medium is treated as three dimensional. The hologram records a three-dimensional interference pattern and therefore is a three-dimensional grating. We have to use a coupled wave theory in order to discuss the characteristics of the volume hologram.

1. Single-Layer Holograms

Single-layer holograms can be absorptive, phase only, or reflective. For an absorptive hologram, we have already calculated the reconstruction efficiency. Since the reconstruction efficiency, η, is proportional to M^2, the maximum value occurs at $M = 1$. This is the case where the object beam intensity, I_O, is equal to the reference beam intensity, I_R. In this case the exposure, E, given in Eq. V-2 varies from zero to twice its dc level. If we assume the recording medium is linear, the corresponding swing of the field transmission, T, is from 1 to 0 with a dc level at $\frac{1}{2}$. That is, for an absorptive hologram of maximum reconstruction efficiency,

$$T = \tfrac{1}{2} + \tfrac{1}{2} \cos (Ky)$$
$$= \tfrac{1}{2} + \tfrac{1}{4}(e^{iKy} + e^{-iKy}). \qquad (V\text{-}20)$$

Hence, a reconstruction wave of unit amplitude will result in a reconstructed wave with an amplitude equal to $\frac{1}{4}$. Thus,

$$\eta_{\max} = (\tfrac{1}{4})^2 = 6.25\%. \qquad (V\text{-}21)$$

A single-layer absorptive hologram therefore has an ultimate reconstruction efficiency of 6.25%.

For a single-layer phase-only hologram, the reconstruction efficiency is given by Eq. V-17 or V-19, and its value is shown in Fig. V-5. The maximum value occurs at $\phi_1 = 1.8$ and is

$$\eta_{\max} = 33.9\% \qquad (V\text{-}22)$$

Hence, the ultimate reconstruction efficiency of a single-layer phase-only hologram is 33.9%.

The single-layer reflective hologram is essentially a single-layer phase-only hologram reconstructed in a reflective manner. The surface relief of the recording medium of a phase-only hologram contains the necessary information that allows the reflective wave from this surface to form a reconstructed image. The efficiency becomes higher if a highly reflective film is evaporated on such a surface. Physically, such a single-layer reflective hologram is a reflective grating. It is well known that a reflective grating can be blazed to obtain very high diffraction efficiency. A single-layer reflective hologram can also have this blazing effect. For example, reflective holograms with reconstruction efficiency of 85% have been reported.[6]

2. Volume Holograms

There are four types of volume holograms: transmitting absorptive holograms, transmitting phase-only holograms, reflective absorptive holograms, and reflective phase-only holograms.

The medium to record a volume hologram must be many wavelengths in thickness. The reference beam and the object beam form a three-dimensional

---

* Of course, if the depth of blazing becomes great enough (as in the case of the 85% efficient hologram), it is not clear that we should call it a "single-layer" hologram. True, it has only a single surface, but contributions to that surface are made from many "layers" (see next section) of the recorded three-dimensional diffraction pattern.

interference pattern which is then recorded as a three-dimensional grating in the medium. The exact behavior of a volume hologram is explained by a coupled wave theory.[7] The details of the theory are beyond the scope of this book, but the results will be described briefly later.

The distinction between the transmitting holograms and the reflective holograms is due to the angle between the reference beam and the object beam.

In the case of a transmitting hologram, both the reference beam and the object beam are introduced from the same side of the recording medium. A simple case of plane wavefronts is shown in Figure V-6a, where the reference

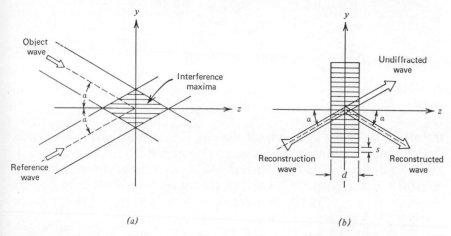

(a)   (b)

Figure V-6 Formation of a transmitting volume hologram using plane reference and object beams incident at equal angles to the normal of the recording medium (a) and the resulting fringe pattern (b).

wave and the object wave are introduced at equal angles from the normal of the recording medium. The interference pattern produced by two such wavefronts contains horizontal strips of maxima at a spacing given by

$$S = \frac{\lambda}{2n \sin(\alpha)}, \tag{V-23}$$

where n is the index of refraction of the recording medium. The fringes recorded in the medium are depicted in Figure V-6b. In the reconstruction process, these fringes act like reflecting layers to the incident reconstruction wave. The situation is very similar to those occuring in X-ray diffractions from crystals. There is interference between the reflected light from different layers. At the angle given by the well known Bragg condition, there is reinforcement, and a bright reconstructed wave is formed. In the case of the transmitting hologram illustrated in Figure V-6a these reflecting layers are perpendicular to the hologram plane. The incident reconstruction wave is reflected toward

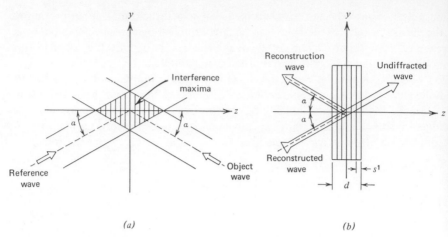

(a)   (b)

Figure V-7   Formation of a reflecting volume hologram (a) and the resulting fringe pattern (b).

the other side of the hologram so that the reconsructed wave appears as a wave transmitted through the hologram (Figure V-6b).

In the case of a reflective hologram, the object beam is introduced almost at 180° from the reference beam. For the simple case of plane wavefronts shown in Figure V-7a, the interference pattern contains vertical strips of maximum at a spacing given by

$$S' = \frac{\lambda}{2n \cos(\alpha)}. \tag{V-24}$$

Hence, the reflecting layers formed in the recording medium arc parallel to the surface of the hologram. In the reconstruction process shown in Figure V-7b, the reflecting layers reflect the incident reconstruction wave back and reinforce at the angle given by the Bragg condition. The reconstructed wave is on the same side of the hologram as the reconstruction wave and thus appears as a wave reflected from the hologram.

If a volume hologram is recorded in an absorptive medium, the interference maxima are recorded as local optical density maxima. The hologram is a three-dimensional optical density grating. Inside the hologram medium, the absorption constant is therefore a function of the spatial coordinates. If the recording medium is again assumed to be linear, the absorption constant varies sinusoidally. It can be written, for example, in the case shown in Figure V-5 as

$$a = a_0 + a_1 \cos\left(\frac{2\pi}{S}y\right), \tag{V-25}$$

where a_0 is the dc absorption constant due to the dc exposure and a_1 is the maximum change due to the sinusoidally varying exposure. Obviously, a_1 is proportional to the depth of modulation, M.

In the case of a phase-only medium, the interference pattern is recorded as the local index of refraction change. For the same example and the same assumption of linear recording, the index of refraction, n, is

$$n = n_0 + n_1 \cos\left(\frac{2\pi}{S} y\right). \tag{V-26}$$

3. Transmitting Absorptive Holograms

The reconstruction efficiency is low because of the absorptive loss of the recording medium; it is given by

$$\eta = e^{-2a_0d/\cos\theta} \sinh^2 \frac{a_1 d}{2\cos\theta}. \tag{V-27}$$

where d is the thickness of the hologram. For a given a_0, the efficiency increases as a_1 (hence the depth of modulation) increases. It reaches a maximum when $a_0 = a_1$ and has a value

$$\eta_{\max} = \frac{1}{27} = 3.7\%. \tag{V-28}$$

η versus a_1 is shown in Figure V-8. The ultimate reconstruction efficiency of a transmitting absorptive volume hologram is 3.7%.

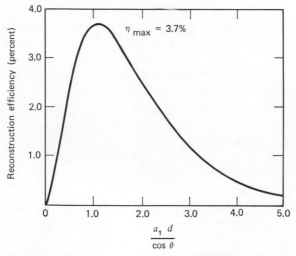

Figure V-8 The variation efficiency, η, varies according to the magnitude, a_1, of the variation of the absorption.

4. Transmitting Phase-Only Holograms

The phase-only recording medium has no absorption, hence, no light energy loss. The theoretical limit of the reconstruction efficiency is therefore expected to be 100%. In fact, the coupled wave theory predicts a reconstruction efficiency

$$\eta = \sin^2 \frac{\pi n_1 d}{\lambda \cos \theta}. \tag{V-29}$$

Hence, the efficiency is an oscillating function of $n_1 d$. The maximum efficiency is 100%. This occurs when $2n_1 d / \lambda \cos \theta$ is an odd integer.

5. Reflective Absorptive Holograms

The reconstruction efficiency of this type of hologram is a complicated function of a_0 and a_1. The coupled wave theory gives an efficiency which is

$$\eta = \frac{a_1{}^2}{\left[2a_0 + \sqrt{4a_0{}^2 - a_1{}^2} \coth\left(\dfrac{d\sqrt{a_0{}^2 - a_1{}^2/4}}{\cos \theta}\right)\right]^2}. \tag{V-30}$$

The efficiency increases monotonically with respect to the depth of the hologram. The maximum efficiency occurs when $a_0 = a_1$. This corresponds to a unit depth of modulation. In this case the efficiency can be calculated from Eq. V-30 by setting $a_0 = a_1$, its value reaches an asymptotic maximum of

$$\eta_{\max} = 7.2\%. \tag{V-31}$$

This is the ultimate reconstruction efficiency of a reflective absorptive volume hologram.

6. Reflective Phase-Only Holograms

The ultimate reconstruction efficiency of this type of hologram is again 100%. As the thickness of the hologram increases, the efficiency increases monotonically and approaches the ultimate limit asymptotically. The explicit expression of the reconstruction efficiency is

$$\eta = \tanh^2 \frac{\pi n_1 d}{\lambda \cos \theta}. \tag{V-32}$$

Figure V-9 shows the efficiency variation as a function of $n_1 d / \cos \theta$.

In conclusion, the above ultimate reconstruction efficiencies are calculated for the holograms with plane waves as reference and object beams. For the holograms with more complex object waves, in general the entire hologram plate will not be exposed to give maximum reconstruction efficiency. We may

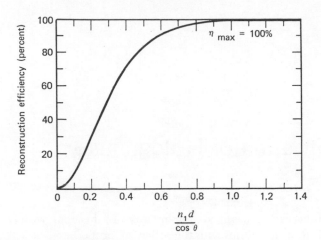

Figure V-9 The reconstruction efficiency, η, varies with $n_1 d/\cos\theta$ for a reflective phase-only hologram.

find many areas on the hologram which do not contribute much to the intensity of the reconstructed wavefront. The efficiency of such a hologram is therefore much lower than the ultimate maximum value given above. Typically, the best efficiency achievable for a complicated object wavefront is about 50% of the best efficiency achievable with a plane object wavefront.

REFERENCES

1. D. H. Close, A. D. Jacobson, J. D. Margerum, R. G. Brault, and F. J. McClung, *Appl. Phys. Letters*, **14**, 159 (1969).
2. J. C. Urbach and R. W. Meier, *Appl. Opt.*, **5**, 666 (1966).
3. T. A. Shankoff and R. K. Curran, *Appl. Phys. Letters*, **13**, 239 (1968).
4. T. A. Shankoff, *Appl. Opt.*, **7**, 2101 (1968).
5. A. A. Friesem, A. Kozma, and G. F. Adams, *Appl. Opt.*, **6**, 851 (1967).
6. N. K. Sheridon, *Appl. Phys. Letters*, **12**, 316 (1968).
7. H. Kogelnik, *Bell Tyst, Tech, J*, **48** 2909 (1969).

VI

Classification of Holograms

A. CATEGORIES

Holograms can be classified in many ways. In this book we have chosen to use classification categories with the following properties:

1. Classification of the hologram within one category does not determine its classification within the other categories.
2. Each hologram has one and only one classification within each category.
3. Each category determines important properties of the hologram.

This listing is given for two reasons. First, the literature abounds in these classifications. To understand current papers in holography, it is necessary to be familiar with the categories normally used. Second, the categories are instructive in that they abstract essential features from complex situations. Comparison of choices within categories is instructive.

The categories we will discuss are the following:

1. Derivation of the reference beam (did the reference beam pass through or reflect off the object or did it bypass the object?).
2. Diffuseness of the object beam (can the object be considered to be a source of diffuse radiation?).
3. Relation of reference beam to object beam (are they colinear or split?).
4. Diffraction region (near field or far field?).
5. Object geometry (natural, focused on hologram, or transformed by lenses or other elements).
6. Recording method (continuously recorded or discretely sampled?).
7. Reconstruction method (transmitted or reflected?).

We will discuss each category separately before we make cross-category comparisons.*

* This discussion is necessarily of limited scope and depth. Many similar discussions can be found in the literature. Perhaps the best is found in reference 1.

B. DERIVATION OF THE REFERENCE BEAM

The reference beam can be derived either from waves directed to the object or from waves not directed to the object. We have called these two cases single-beam holography and split-beam holography, respectively. Chapters III and IV describe these two cases in detail. The single-beam holograms require no *a proiri* knowledge of the object distance and, hence, are easier to use outside the laboratory. On the other hand, the split-beam hologram images are easier to reconstruct than single-beam holograms because the image can be broken away from the reconstructing beam.

C. DIFFUSENESS OF THE OBJECT BEAM

A diffuse object beam can come about in either of two ways. If a diffuser is inserted between the source and the object, even a nondiffuse object will produce a diffuse object beam. If the object itself is diffusing, even a non-diffuse wavefront incident on that object will produce a diffuse object beam. In either case we assume:

1. Each point on the object contributes wave information to each point on the hologram surface and
2. none of the path differences thus introduced are great enough to affect the coherence between the reference beam at any point and the object beam (from the whole object) at that point.
Under these conditions the hologram has a number of unique features.

First, each "point" on the hologram contains information about the whole object. In fact, the information which can be obtained by reconstructing from any particular point on the hologram is precisely the information that could be obtained from viewing the original object through an aperture (of the size and location of the reconstruction area) in an otherwise opaque screen. All the rules of normal image formation apply, so the resolution improves and the depth of field decreases as the reconstructing area increases.[2]

Second, the diffuse object beam allows the hologram exposure to be spatially uniform. This is in contrast to the case of ordinary photography, in which the exposure is "proportional" (see Section I-B-2) to the intensity in the image and, hence, very nonuniform. Thus, while in ordinary photography optimum exposure can occur for only one part of the object (usually the brightest part), in diffuse-beam holography optimum exposure can occur for all parts of the object. This increase in dynamic range was first pointed out by Leith and Upatnieks.[3] Clearly the dynamic range is limited only by noise factors and, hence, is essentially unlimited. Numerous other advantages of diffuse illumination have to do with the accuracy and signal-to-noise ratio of the reconstructed image.[3]

Third, transparent objects, which would be very difficult to view directly without the diffuser (you see the point light source through the image of the transparency), appear to be floating in air if they are placed very close to a diffuser (you see the diffuser through the transparency). This is not to say that the image of a transparency illuminated with nondiffuse light cannot be viewed, a microscope with very limited depth of field makes the reconstructed image quite viewable. Alternatively, a diffusing screen in the real image plane of the nondiffuse-beam hologram also produces an easily viewed image.

Fourth, diffuse object beams produce speckled images. Speckling is an inevitable result of imaging an object with random phase variations with coherent waves.[4] The speckled pattern (Figure VI-1) arises from local phase

Figure VI-1   The pattern shown here is a typical speckle pattern which arises from random phase modulation of a coherent wavefront.

variations which average (over a large enough region) to zero, but vary locally over the full range from 0 to 2π. Several ways to eliminate this unpleasant effect have been used. The diffused object beam has been changed in time to produce an ensemble average of zero (in analogy to the spatial average of zero discussed above). A large-aperture imaging system (replacing the 5 to 7 mm aperture of the human eye) produces a large area for spatial averaging.[5]

Diffuseness of the reference beam is not a very useful category for two reasons. The most useful reference beams are point-source beams because they are easiest to reproduce. We assume that the point source radiates in all directions of interest for our hologram, but this is not a very useful definition of diffuseness. If, for some reason, we went to an extended reference-beam source, diffuseness of that source would again be of interest only to the extent that it is necessary to achieve radiation to all points of the hologram.[6,7]

D. RELATION OF REFERENCE BEAM TO OBJECT BEAM

The case in which the reference beam is colinear with the object beam is so different from the case of noncolinear beams as to be worthy of special mention. To two cases are often called " on-axis " and " off-axis." The off-axis hologram is also called a "sideband" hologram in analogy to the radio terminology. Gabor's first holograms (Section IV-B) were on-axis. This required filtering out the undiffracted beam and also led to interference between the real and virtual images. Stroke[8] has pointed out that the images can be viewed from off-axis in certain cases. The off-axis holograms can produce full separation of the image beams from each other and from the undiffracted beam.

E. DIFFRACTION REGION

The nature of the wavefront varies with the distance, z, between the object and the observation point. Observed at small z, the pattern varies rapidly with z and consists primarily of alternating light and dark patterns the same shape as the object and lying within the geometrical projection of the object. Figure VI-2 shows photographs of the diffracted wavefront produced by a cookie cutter and recorded at various values of z. This near-field phenomenon is often called Fresnel diffraction. Eventually, as z increases, the diffraction pattern becomes simplified. In the far field (Fraunhofer diffraction region), the pattern grows in size as z increases, but the basic pattern is unchanged.

Figure VI-3 shows the far-field pattern produced by a cookie cutter pattern at various values of z. In the far field the interesting " structure " appears outside the projected outline of the object. This is in contrast to the near field where the interesting structure is inside the projected outline. Caulfield and Eden[9] have treated the simplest case, the circular aperture, in some detail.

The literature of holography abounds in mistatements of the conditions for near-field (Fresnel) and far-field (Fraunhofer) diffraction. The proper statement is

Figure VI-2 The Fresnel diffraction pattern varies in detail as the distance from the object increases.

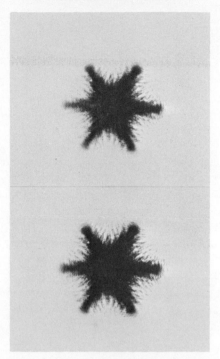

Figure VI-3 The Fraunhofer diffraction pattern changes only in size as the distance from the object increases.

$$z \ll z_0 \Rightarrow \text{near-field, Fresnel}$$

$$z \gg z_0 \Rightarrow \text{far-field, Fraunhofer,}$$

where

$$z_0 = \frac{k\lambda^2}{a}.$$

Here λ is the wavelength, a is a charactcristic halfwidth of the object (e.g., a is the radius of a circular aperture), and

$k = 1$ for incident plane waves.

$k > 1$ for divergent waves, and

$k < 1$ for convergent waves.

Omission of the factor k (which can be any positive number) can lead to highly inaccurate calculations. The mathematical distinction between the two cases is discussed in Section II-C-1.

Mittra and Ransom[10] have examined holographic imaging in the two cases (Fresnel and Fraunhofer) in mathematical detail. They predict that holograms recorded in the Fresnel region will reconstruct images which propogate as the original wavefront both forward in time (virtual image) and backward in time (real image). On the other hand, a Fraunhofer hologram may not contain enough information to reconstruct the Fresnel region. This has been confirmed by experiment, suggesting some very important differences between Fresnel and Fraunhofer holograms.

It is useful to approach Fresnel and Fraunhofer holograms in terms of their resolution properties.* A useful way to discuss resolution is to relate the space–bandwidth product, $(SW)_O$, of the object to the space–bandwidth product, $(SW)_H$, of the hologram recording medium. The space–bandwidth product is a measure of the information content of a signal (in the intuitive sense of information). Here S, the space, is the area over which the information is distributed and W, the two-dimensional spatial frequency width, is the inverse of the smallest resolution area. Thus the two-dimensional space-bandwidth product is

$$SW = \frac{\text{area of image}}{\text{area of smallest cell}}$$

One of the best discussions of the space-bandwidth product has been given by Lohmann.[11] Unfortunately, Lohmann's paper is not in the open literature. Much of this material has been covered by De Velis and Reynolds[12] and by Stroke.[13] For reference, Lohmann listed the following order-of-magnitude SW values:

Medium	SW
Microscope, TV	10^5
Good photo lens	10^7
Long focal length lens used for aerial reconnaissance	10^8
100-cm² Kodak 649F film	10^{10}

An object of size $\Delta x \, \Delta y$ and minimum point size $\delta_x \, \delta_y$ has

$$(SW)_0 = \frac{\Delta x \, \Delta y}{\delta_x \, \delta_y}. \tag{VI-1}$$

---

* Here the hologram is called a Fraunhofer hologram if each point of the object contributes to each point of the hologram. Otherwise, it is called a Fresnel hologram.

By the sampling theorem (see Section VI-G), the image in spatial frequency coordinates is built up of points separated by $\delta v_x = 1/\Delta x$ and $\delta v_y = 1/\Delta y$. The total ranges are $\Delta v_x = 1/\delta x$ and $\Delta v_y = 1/\delta_y$. Thus, the image must have

$$(SW)_H = \Delta v_x \, \Delta v_y / \delta v_x \, \delta v_y$$

$$= \frac{\Delta x \, \Delta y}{\delta_x \, \delta_y}$$

$$= (SW)_0. \qquad \text{(VI-2)}$$

For perfect economy we would want

$$(SW)_H = (SW)_0. \qquad \text{(VI-3)}$$

The discussions by the various authors differ somewhat at this point and some confusion has arisen. It suffices to note two sources of this confusion. First, it is seldom stated whether we are talking about two-dimensional or one-dimensional SW's. For the perfectly symmetrical case, of course,

$$(SW)_{2-D} = (SW)^2_{1-D}. \qquad \text{(VI-4)}$$

Second, the resolution cell size is not defined in the same way by all authors. The general results are all that need interest us here. For all types of holograms except the off-axis Fresnel hologram.

$$(SW)_H \gtrsim (SW)_0 ; \qquad \text{(VI-5)}$$

but for the off-axis Fresnel hologram

$$(SW)_H \gg (SW)_0. \qquad \text{(VI-6)}$$

The problem with the off-axis Fresnel hologram is the quadratic fringe pattern problem discussed in Section VI-F. The off-axis Fresnel hologram does not make efficient use of the SW of the recording medium. On the other hand, the off-axis Fresnel hologram is the easiest to make and the easiest to reconstruct. It gives the best three-dimensional pictures.

In connection with the space–bandwidth product, it is worthwhile to note that the one-dimensional $(SW)_H$ is the number of linear fringes recorded in that direction.

F. OBJECT GEOMETRY

We refer to the object geometry as "natural" if no optical elements intervene between the object and the hologram. Clearly, this need not be the case. The simplest optical element which can be inserted is a lens or lens system. Two special cases of intervening lens systems are of interest. First, the case in which the object is in one focal plane and the hologram is in the other focal plane leads to the Fourier transform hologram. Second, the case in which the object is imaged onto the hologram plane leads to the focused-image hologram.

A lens converts a complex object distribution $f(x, y)$ in one focal plane into its Fourier transform $F(U, V)$ in the other focal plane by a process described in Section II-B. Note that since the object appears to be at infinity, the Fourier transform hologram is a Fraunhofer hologram. Another Fourier transform can now be performed by a second lens. The net effect of two successive Fourier transforms is to restore the original wavefront, except that the signs of the coordinates are reversed, that is, the image is upside down and back-

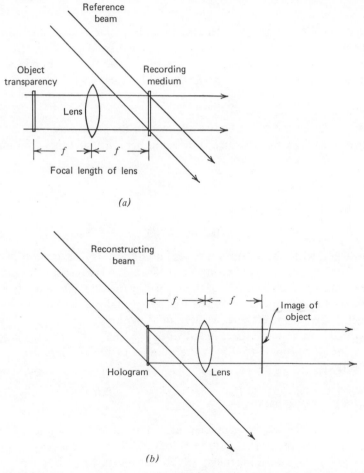

(a)

(b)

Figure VI-4 The formation *(a)* and reconstruction *(b)* processes of Fourier transform holography using a lens to perform the transformation.

ward. This same operation converts the conjugate image (which was already reversed) into a copy of the original wavefront. Therefore, the Fourier transform hologram can yield viewable images (both real, but one reversed in direction) if the reconstructed wavefront is transformed by a lens. The two steps (construction and reconstruction) are shown in Figure VI-4.

It is possible to form a Fourier transform hologram without using a lens by a process suggested by Stroke[13] and illustrated in Figure VI-5. A point

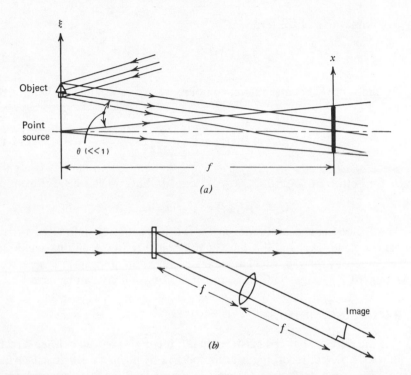

Figure VI-5 The formation (*a*) and reconstruction (*b*) processes for lensless Fourier transform holography.

source for the reference beam is placed in the plane of the object and displaced from the object. The field from the object is

$$U_O(x) = \int_{\xi_1}^{\xi_2} T(\xi) \exp\left[i\left(\frac{k}{2f}\right)(x - \xi)^2\right] d\xi, \qquad (VI-7)$$

where $T(\xi)$ is the object transmission (or reflection). If we assume small objects and relatively large holograms,

$$(x - \xi)^2 \cong x^2 - 2x\xi. \qquad (VI-8)$$

Therefore,

$$U_O(x) = \exp\left[i\left(\frac{k}{2f}\right)x^2\right] \int_{+\xi_1}^{\xi_2} T(\xi) \exp\left[-i\left(\frac{k}{f}\right)x\xi\right] d\xi. \qquad \text{(VI-9)}$$

That is, the object field differs from the Fourier transform

$$U_O{'}(x) = \int_{\xi_1}^{\xi_2} T(\xi) \exp\left[-i\left(\frac{k}{f}\right)x\xi\right] d\xi \qquad \text{(VI-10)}$$

by a multiplicative phase term

$$\exp\left[i\left(\frac{k}{2f}\right)x^2\right]. \qquad \text{(VI-11)}$$

By choosing a coplaner reference point, we assure a reference field of the form

$$U_R(x) = U_{R_0} \exp\left[i\left(\frac{k}{2f}\right)x^2\right]. \qquad \text{(VI-12)}$$

Thus, for both $U_R U_O{}^*$ and $U_R{}^* U_O$, the quadratic term cancels, for example

$$U_R{}^*(x) U_O(x) = U_R{}^*(x)_0 U_O{'}(x). \qquad \text{(VI-13)}$$

The focused-image hologram arises when the object is focused on the hologram plane. Perhaps the most significant advantage of this technique is that, in reconstruction, small angular deviations do not lead to large spatial deviations. Thus, large, wide-band reconstruction sources can be used.[14]

G. RECORDING METHOD

The interference pattern exists in space. It can be recorded either directly or indirectly. Direct recording is accomplished by placing a sensitized surface directly into the interference pattern. Indirect recording is accomplished by sampling the interference pattern at a number of discrete points and producing an optical transparency with discrete properties corresponding to the positions and intensities measured. For both direct and indirect recording, the hologram is a collection of discrete points. In the direct recording case the sampling is provided by the graininess of recording medium. In indirect recording, the sampling is inherent and deliberate. In both cases the question must be asked: "How many sampling points must be used in order to assure a good hologram?"

The first step in answering that question is to relate sampling to the space-bandwidth product SW of the object. The classical sampling theorem, given in one dimension for simplicity, is easy to derive. We suppose that the function we wish to sample is $f(x)$. Then we know that if $f(x)$ has nonzero spatial frequencies v only for $|v| \leq v_0/2$, we can expand $f(x)$ in terms of the complete orthonormal set $\{\phi_n(x)\}$ where

$$\phi_n(x) = \sqrt{v_0} \; \text{sinc}\left[\pi v_0\left(x - \frac{n}{v_0}\right)\right], \qquad \text{(VI-14)}$$

where we use the standard notation

$$\text{sinc } y \triangleq \frac{\sin y}{y}. \qquad \text{(VI-15)}$$

This particular set of orthonormal functions is interesting because its expansion coefficients are easy to compute. We remember that by orthonormality we mean

$$\int_{D_x} \phi_n(x)\phi_m^*(x)\, dx = \delta_{n,m}, \qquad \text{(VI-16)}$$

where D_x is the domain of definition of $\{\phi_n(x)\}$ and of $f(x)$ and $\delta_{n,m} = 1$ if $n = m$ and of $\delta_{n,m} = 0$ if $n \neq m$.
Clearly, if we write

$$f(x) = \sum_n c_n \phi_n(x), \qquad \text{(VI-17)}$$

then

$$\int_{D_x} f(x)\phi_n^*(x)\, dx = \sum_m c_m \int_{D_x} \phi_m(x)\phi_n^*(x)\, dx$$
$$= \sum_m c_m \delta_{mn} = c_n. \qquad \text{(VI-18)}$$

Applying Eq. VI-18 to Eq. VI-14, we arrive at the classical sampling theorem

$$f(x) = \sum_{n=-\infty}^{\infty} f\left(\frac{n}{v_0}\right) \text{sinc}\left[\pi v_0\left(x - \frac{n}{v_0}\right)\right]. \qquad \text{(VI-19)}$$

Finally, if $f(x) = 0$ for $|x| > x_0/2$ and $N \triangleq v_0 x_0$ then

$$f(x) = \sum_{-N/2}^{N/2} f\left(\frac{n}{v_0}\right) \text{sinc}\left[\pi v_0\left(x - \frac{n}{v_0}\right)\right]. \qquad \text{(VI-20)}$$

Thus, for a function with $SW = x_0 v_0 = N$, we can reproduce the function exactly from samples of the function taken at N evenly spaced intervals (spacing $= 1/v_0$).

It often happens that N is prohibitively large. In this case undersampling must be used. If we stick to evenly spaced sampling, we can spread the samples across the full D_x and achieve high resolution with lowered fidelity ("aliasing") or sample a small part of D_x and get full fidelity. More exotic means for under-sampling are discussed in Chapter XVI-E. These offer hope for good holograms with orders of magnitude fewer samples than N.

REFERENCES

1. F. Bestenreiner and R. Deml, *Z. Angew. Phys.*, **25**, 243 (1968).
2. E. N. Leith and J. Upatnieks, *Sci. Amer.* **212**, (6), 24 (1965).
3. E. N. Leith and J. Upatnieks, *J. Opt. Soc. Amer.*, **54**, 1295 (1964).
4. L. I. Goldfischer, *J. Opt. Soc. Amer.*, **55**, 247 (1965).
5. E. N. Leith and J. Upatnieks, *Appl. Opt.*, **7**, 2075 (1968).
6. H. J. Caulfield, *Phys. Letters*, **27A**, 319 (1968).
7. G. B. Brandt, *Appl. Opt.* **8**, 1421 (1969).
8. G. W. Stroke, D. Brumm, A. Funkhouser, A. Labeyrie, and R. C. Restrick, *Brit. J. Appl. Phys.*, **17**, 497 (1966).
9. H. J. Caulfield and D. D. Eden, *Amer., J. Phys.*, **34**, 439 (1966).
10. R. Mittra and P. L. Ransom, in Proceedings *Symposium on Modern Optics*, Polytechnic Institute of Brooklyn, 1967, p. 619.
11. A. W. Lohmann, "The Space Bandwidth Product Applied to Spatial Filtering and to Holography," IBM Research Paper RJ-438, 1967.
12. J. B. De Velis and G. O. Reynolds, *Theory and Applications of Holography*, Addison-Wesley, Reading, Mass., 1967.
13. G. W. Stroke, *An Introduction to Coherent Optics and Holography*, Academic Press, New York, 1966.
14. L. Rosen, Appl. Phys. Letters, **9**, 337 (1966).

VII

Limitations of Holography

A. INTRODUCTION

In principle, there are only two types of limitations—incurable and curable. There is great danger in classifying a limitation as incurable. The ground rules change as progress is made. Thus, we will try to note which of the limitations seem likely to be overcome, which will require a whole new approach, and which are inherent. Many of the limitations are not profound or surprising, but they are also seldom displayed explicitly.

B. OBJECT MOTION

Motion of the object can have two types of effects on the hologram. First, it can change the degree of coherence between the object wavefront and the reference wavefront. Second, it can effectively replace the instantaneous object wavefront with a time-averaged wavefront. The general theory of holograms with moving objects has been given by Goodman.[1] It will not be repeated here. Rather we will offer several basic observations.

1. It is possible to move the object in such a way that the interference pattern between the object wavefront and the reference wavefront is very insensitive to the motion. Neumann[2] has shown how to calculate the proper direction of motion.

2. If the exposure time is short enough, the motion will have a negligible effect. As expected from Neumann's work, the direction of motion is very important. A very simple analysis for a moving point scatterer was given by Brooks et al.[3] based on Figure VII-1, where θ is the angle of motion relative to the illuminating ray, Δ is the amount of motion during the exposure, and α is the angle between a point on the hologram and the illuminating ray. They confine consideration to the case in which θ and α are coplanar, which is the most sensitive case. The value of Δ which leads to a phase shift of 2π at the hologram (and hence complete "washing out" of the hologram) is

$$\Delta = \left| \frac{\lambda}{\cos \theta - \cos (\theta - \alpha)} \right|. \qquad \text{(VII-1)}$$

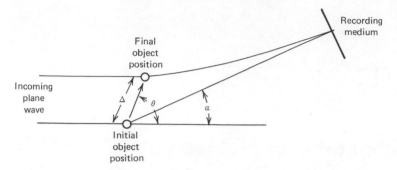

Figure VII-1 A diagram useful for discussing object motion effects in holography based on the pioneering work of Brooks, Heflinger, and Wuerker in *IEEE J. Quantum Electronics QE-2*, 275 (1966).

For $\theta = \pm\pi/2$,

$$\Delta = \left| \frac{\lambda}{\sin \alpha} \right|. \tag{VII-2}$$

We can make $|\sin \alpha|$ very small, so large motions can be tolerated. Similarly, if $\theta = 0$

$$\Delta = \left| \frac{\lambda}{(1 - \cos \alpha)} \right|. \tag{VII-3}$$

For $|\alpha| > \pi/2$ (front-lighted object), $\cos \alpha < 0$ and, therefore, $\Delta < \lambda$ in Eq. VII-3, and great motion sensitivity occurs. The restriction of motion to distances $\ll \Delta$ can be very restrictive indeed. Consider a typical case of $\Delta = 2\lambda(\theta = \pi/2, \alpha = \pi/6)$. For velocity V (cm/sec), the exposure time, t, must be

$$t \ll \frac{\lambda}{v}. \tag{VII-4}$$

A man can run at about 10^3 cm/sec (not very fast as real objects go). A typical visible wavelength is $\lambda = 0.5 \times 10^{-4}$cm. Thus, we would require $t \ll 0.5 \times 10^{-7}$ sec to record a man at $\theta = \pi/2$ and $\alpha = \pi/6$. This requires a Q-switched laser or an electrooptic shutter. The energy supplied to the recording medium during the pulse must be sufficient to record the hologram (see Section V-C). The coherence of ordinary Q-switched lasers is usually not great enough to permit holography, but restrictions on the effective cavity are sufficient to give great coherence.[4-6]

3. The extreme sensitivity of a front-lighted object to blurring of the hologram fringes has been used to study object motions. This important application is discussed in detail in Chapter XII.

4. The time-averaged object wavefront can be used to reconstruct a wave-front of a nonexistent object. This is discussed in Section X-B.

5. The effect of motion of the object on the image can be written as a time-independent function. The transfer function can be generated holographically if the motion is known, so an image of the stationary object can be reconstructed from a hologram taken while the object is moving. See Section X-F for similar effects.

C. RECORDING MEDIUM MOTION

The restrictions on recording media are very easy to discuss. The motion must be much less than the resolution size required for recording the hologram (see Section V-B). Smith[7] has treated this problem in detail. As an example of how great the toleration to recording medium motion can be, Stroke[8] took a hologram of a stationary object with a *cw* (continuous wave) laser and hand-held film.

Of course, the recording medium can be moved purposely to compensate for object motion or to generate a synthetic image (Sections VII-B and X-B).

D. OBJECT SIZE

The size and location of the objects which can be recorded by holography is limited by the coherence distance of the object for any particular experimental arrangement.

Once the geometry of the hologram arrangement is fixed and the coherence length of the source is given, a volume in space is defined such that waves can travel to any object in that volume and then to the recording medium and still have traveled a path distance different from the path distance of the reference waves (both measured from the wave source) by no more than the coherence distance. We can call that volume in space the "coherence volume." Anything within the coherence volume will be imaged and anything outside the coherence volume will not be imaged.*

Changing the physical arrangement can change the shape and location of the coherence volume.[11] Unfortunately, this is inconvenient if we do not know ahead of time where the object is going to be or we do not want to fill the region with mirrors and beam splitters. Coherence extension[4-6] changes the dimensions of the coherence volume. In LRB holography (Section IV-C) the coherence volume includes all space which can send waves into the recording apparatus.

---

* Actually, there can be imaging outside the coherence volume. This subject is not complicated and is well established.[8-10] It is omitted here for brevity. This reflects no judgement on its potential importance.

Like most of the limitations of holography, the finite coherence volume of split-beam holography can become an advantage in special circumstances. In Section VIII-F, we discuss the usefulness of the finite coherence volume in fog penetration.

E. RECONSTRUCTING SOURCE

The requirement on the reconstructing source is simple to state: "the reconstructing source should duplicate the reference beam." Any deviation from duplication causes errors and distortions as noted in Section III-D. Fortunately, however, the errors and distortions are seldom bothersome if all we wish to do is reconstruct a three-dimensional image of an object. It is usually sufficient to use an incoherent light source, a pinhole to simulate a point source, and a color filter. Inexpensive hologram viewers without lasers can be built.

For some cases (e.g., electron holograms, X-ray holograms, and LRB holograms), the reference beam cannot be derived from a point source. The finite extension of the point source leads to limited resolution in the reconstructed image if the normal reconstruction techniques (using a point-source reconstruction beam) are employed. Stroke et al.[12] showed that this lost resolution could be retrieved by using a reconstruction beam source more nearly like the reference beam source. When this is possible, resolution can be improved significantly.

F. HOLOGRAM WAVE SOURCES

The question to be answered here is: "How coherent must our wave source be?" Unfortunately, that question is still too poorly defined to be answered.

It is convenient to divide coherence (see Section II-B) into two types—"spatial" and "temporal."

Spatial coherence is concerned with the spatial separation between two points which can be tolerated without destroying the coherence in the far field of the two points. It is intimately related to the size of the wave source. The smaller the source, the greater the spatial coherence. The requirements on the source extension, Δx_s, are set by the number of fringes, N, we wish to record or, equivalently, on the space-bandwidth product we demand. Even then the answer is complicated. If we wish to illuminate an object with dimensions Δx, the number of fringes, N, that can be recorded on the hologram is $\Delta x_0/\delta x_0$, where δx_0 is the resolution limit for the system. For wavelength λ and a Fourier-transforming lens of length f,

$$N \leq \frac{\Delta x_0{}^2}{\lambda f}. \tag{VII-5}$$

The spatial coherence is approximately equal to

$$\Delta x_c = \frac{\lambda}{\alpha_s}, \tag{VII-6}$$

where α_s is the angular size of the source.[13] Thus, so long as the source dimensions and size are such that

$$\Delta x_c > \Delta x_0, \tag{VII-7}$$

the spatial coherence does not reduce N.

Temporal coherence is concerned with the spread in wavelengths, $\Delta\lambda$. To resolve N fringes there must be a difference in optical path lengths of $\frac{1}{2}N\lambda$.

The coherence distance, ℓ_c, is the path difference which results in a 2π phase shift (complete "washing out" of the interference pattern). Therefore, we must require

$$\frac{1}{2} N\lambda < \ell_c = \frac{\lambda^2}{\Delta\lambda}. \tag{VII-8}$$

A more convenient expression derived from Eq. VIII-8 is

$$\frac{\lambda}{\Delta\lambda} > \frac{1}{2} N. \tag{VII-9}$$

Clearly if we are content with relatively small N, we can use holography without lasers or other highly coherent sources. Interference filters with $\Delta\lambda = 10$ Å are available commercially, so N values of several hundred can be obtained with arc lamps. As early as 1965 Lohmann[14] and Mertz[15] showed that spatial coherence across the whole hologram was not necessary. These facts have led to considerable investigation of so-called "incoherent" holography. Of course, "incoherent" holography preceded "coherent" holography by 15 years. Only recently, however, have high quality images been produced in this way.[16-18]

G. COPYING

Holograms can be copied for any one of several reasons. First, multiple copies are useful for mass markets or for uses which might risk damage to the original. Second, copying is used to change from one recording medium to another. This often occurs if the original recording was made on fast film to allow motion stopping, but a more grainless medium is desired. Third, copying can be used to change the hologram to a different form or to record a processed image.

The most straightforward approach to copying a hologram is to copy it like any other transparency by projection printing or contact printing when the resolution size is large enough. What "large enough" means depends on how good the contact printer is. Good contact printers can resolve structures as small as a few microns. The important thing is to assure large enough spacing by choosing the proper hologram recording method. Let us suppose the original hologram was recorded with $\gamma = \gamma_1$. Then, rewriting Eq. V-4, the original hologram has transmission*

$$T_1 \propto 1 - \frac{\gamma_1}{2} \frac{E_R E_O^* + E_R^* E_O}{|E_O|^2 + |E_{R/}|^2}. \qquad \text{(VII-10)}$$

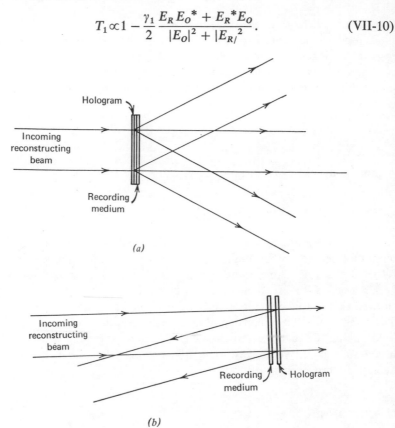

(a)

(b)

Figure VII-2 The experimental setup for copying (a) a transmissive hologram. (b) a reflective hologram using the undeviated beam as a reference beam.

* We have chosen to use the photographic γ (Sec. VB1) rather than the transfer curve slope β (Sec. VB2) because γ is the quantity usually tabulated by photographic suppliers.

If the copy is made with $\gamma = \gamma_2$, we can show that the transmission of the copy is

$$T_2 \propto |T_1{}^2|^{-\gamma_2/2} = 1 + \frac{\gamma_1 \gamma_2}{2} \frac{E_R{}^* E_O + E_R E_O{}^*}{|E_R|^2 + |E_O|^2}. \qquad \text{(VII-11)}$$

Thus, the effect is as though the original hologram had been made with $\gamma = -\gamma_1 \gamma_2$. The positive–negative reversal does not affect the diffraction (see Section I-C). If $|\gamma_2| > 1$, then $|\gamma_1 \gamma_2| > |\gamma_1|$ and the reconstruction efficiency can be improved.

Another way to copy a hologram is to reconstruct the image to serve as the object beam for a new hologram. The reference beam can be supplied in any of many ways. The easiest way is to use the undeviated beam as the reference beam. The experimental setup is shown in Figure VII-2. This technique is applicable in situations where direct copying is not, because the method itself imposes no resolution limits. It is easy to show that, to first order in all of the small quantities, Eq. VII-11 applies to this type of copying also.

REFERENCES

1. J. W. Goodman, *Appl. Opt.*, **6**, 857 (1967).

2. D. B. Neumann, *J. Opt. Soc. Amer.*, **58**, 447 (1968).

3. R. E. Brooks, L. O. Heflinger, and R. F. Wuerker, *IEEE J. Quantum Electron.*, **2**, 275 (1966).

4. F. J. McClung and D. Weiner, *IEEE J. Quantum Electron.* **1**, 94 (1965).

5. A. D. Jacobson and F. J. McClung, *Appl. Opt.*, **4**, 1509 (1965).

6. L. D. Siebert, *Appl. Phys. Letters*, **11**, 326 (1967).

7. H. M. Smith, *Principles of Holography*, Wiley, New York, 1969.

8. G. W. Stroke, A. Funkhouser, C. Leonard, G. Indebetouw, and R. G. Zech, *J. Opt. Soc. Amer.*, **57**, 110 (1967).

9. M. Born and E. Wolf, *Principles of Optics*, Pergamon Press, Oxford, 1964.

10. J. B. Develis and G. O. Reynolds, *Theory and Applications of Holography*, Addison-Wesley, Reading, Mass., 1967.

11. H. J. Caulfield, *J. Opt. Soc. Amer.* **58**, 276 (1968).

12. G. W. Stroke, R. Restrick, A. Funkhouser, and D. Brumm, *Phys. Letters*, **18**, 274 (1965).

13. G. W. Stroke, *An Introduction to Coherent Optics and Holography*, Academic Press, New York, 1966.

14. A. W. Lohmann, *J. Opt. Soc. Amer.*, **55**, 1555 (1965).

15. L. Mertz, *Transformations in Optics*, Wiley, New York, 1965.

16. O. Bryngdahl and A. Lohmann, *J. Opt. Soc. Amer.*, **58**, 625 (1968).

17. J. Upatnieks and E. N. Leith, *J. Opt. Soc. Amer.*, **57**, 563 (1967).

18. M. Kato and T. Suzuki, *J. Opt. Soc. Amer.*, **59**, 303 (1969).

VIII

Applications to 3-D Photography

A. INTRODUCTION

The following discussion is concerned with a three-dimensional object whose image we wish to record. We will not consider the problem of magnifying the image or the problem of studying image motion in this chapter.

B. EXTRA-INFORMATION HOLOGRAPHY

Holography normally records the intensity and phase of a single polarization state at a single wavelength. Multiple holograms offer the possibility of recapturing the missing information.

Color holography requires three holograms (one in each of three colors). The three holograms, each reconstructed with its own wavelength, must be so arranged that their images overlap. It will appear below that the three holograms can be recorded on one photographic plate, provided that intermodulation problems can be minimized

Polarization holography requires two holograms, one with each of two orthogonal polarization states. The reconstructed images must be of the proper polarization and must be precisely aligned and coherent, and they must accurately reflect the original relative intensities. For these reasons polarization holography is even more difficult than color holography.

C. COLOR HOLOGRAPHY

The first approach to color holography was the multiple-exposure technique of Leith and Upatnieks.[1] Each exposure was made with a different laser, using the reference beam direction to key each hologram. The directions were widely separated to minimize cross talk. The technique works, although it is difficult and cumbersome and leads to noise in many directions.

The next key development was the work by Denisyuk,[2] who adapted the Nobel prize winning color photography technique of Lippman[3] to holography. Basically, the idea is that the light can be made to set up standing waves

70

within the recording medium. The positions of the interference maxima correspond to the three-dimensional grating needed to convert the reference wavefront into the object wavefront. The key observation by Lippman was that these "three-dimensional gratings" are highly wavelength selective. A Lippman plate reflects little except at the constructing wavelength. On transmission, it modulates most strongly the construction wavelength. This then has become the basis for two types of color holography. The first type[4,5] uses the Lippman technique to encode the various monochrome holograms with multiple reference beam directions. The holograms are then reconstructed with the constructing laser. This provides independent control of the intensities of the three colors to insure accurate color rendition. The second type of color holography[6-9] makes use of a reference beam incident from the rear (i.e., from the opposite side from the object beam). In this case the Lippman plate becomes a reflection-reconstructed hologram. When illuminated with a white light (usually a zirconium arc lamp), the hologram reflects only the constructing wavelengths. This method requires careful control of the emulsion thickness[4,9] (which governs the color reflected) and careful control of the relative exposure of the three holograms. Note that this method does not allow *a posteori* control of the color balance. Fortunately, the human observer is very insensitive to color inaccuracies in familiar scenes.

D. POLARIZATION HOLOGRAPHY

Two types of polarization holography must be distinguished. The earlier type[10] took advantage of the fact (Section II-C) that only that polarization component of the object beam which is polarized identically with the reference beam can produce an interference pattern. The later type, first suggested by Lohmann,[11] required two orthogonally polarized reference beams to record the orthogonal polarization states of the object and reconstruct them in phase as described in Section VIII-B.

Recording of a single polarization state is generally the case in holography since most lasers are polarized. Most objects are depolarizing,[12] but some interesting polarizing effects (known to photographers for many years) can be recorded in this way. When the object has small ridges or cracks, for example, fingers and cracked integrated circuits, polarization analysis is very valuable. If the object is birefringent, its effects on polarization can be studied holographically.

The full polarization recording suggested by Lohmann[11] has been refined a great deal since then. Recent work by Fourney et al.[13] is shown in Figure VIII-1. In the figure the object and its holographic image are compared for a variety of analyzer positions.

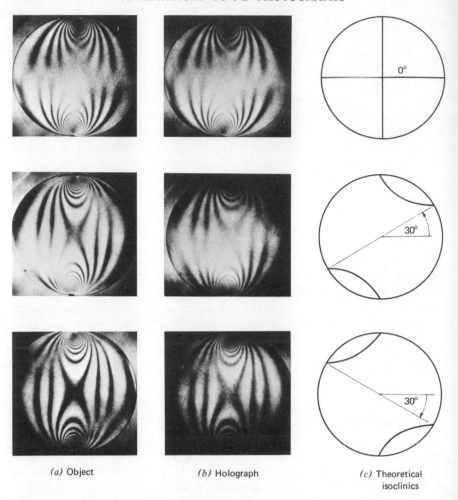

(a) Object   (b) Holograph   (c) Theoretical
                                                                isoclinics

Figure VIII-1 A comparison of polarization effects in the original object with polarization effects in an image of the object produced by holography. (a) Object, (b) holograph, and (c) theoretical isoclinics. From Founey, Waggoner, and Mate, *J. Opt. Soc. Am.*, **58**, 589 (1966).

E. DISTORTION REDUCTION

The image formed by an optical system is (by the linearity property discussed in Section II-A) the sum of the contributions from each point on the object. If a point of unit magnitude at position x_0 in the object plane produces response $t(x_i, x_0)$ at position x_i in the image plane, then an object wave $U(x_0)$ would produce an image

$$V(x_i) = \int_{object} U(x_0)t(x_i, x_0) \, dx_0 . \qquad \text{(VIII-1)}$$

The function $t(x_i, x_0)$ is the "point transfer function" of the system. Suppose the point transfer function is not the diffraction-limited form $t_0(x_i - x_0)$, but instead some value $t(x_i, x_0)$. Then if we could multiply the incoming wave by $t_0(x_i - x_0)/t(x_i, x_0)$, we would have

$$V_0(x_i) = \int_{object} U(x_0)t(x_i, x_0) \frac{t_0(x_i - x_0)}{t(x_i, x_0)} \, dx_0$$

$$= \int_{object} U(x_0)t_0(x_i - x_0) \, dx_0 . \qquad \text{(VIII-2)}$$

A hologram of $t_0(x_i - x_0)/t(x_i, x_0)$ is, by definition, a hologram such that the imperfect point transfer function wavefront is transformed into the diffraction-limited point transfer wavefront. That is, it is a hologram made by the interference between the wavefront from the given lens and the wavefront from a diffraction-limited lens.

This idea of Upatnieks et al.[14] allows compensation for the lens so that good images can be formed. If the compensation is not made, then the holographic image can be deconvoluted. That is, the wavefront $V(x_i)$ can be Fourier transformed to give

$$\mathcal{F}[V(x_0)] \cdot \mathcal{F}[t(x_i, x_0)].$$

We can multiply this by the function

$$\frac{\mathcal{F}[t_0(x_i - x_0)]}{\mathcal{F}[t(x_i, x_0)]}$$

which can be gotten from a Fourier transform hologram using the given lens and a diffraction-limited lens. Retransforming leads to $V_0(x_i)$. If, however, only the squared magnitude of the image is recorded, as in photography, the complex information is lost forever and this process can never restore the image. Nevertheless, a great deal of the information can be restored even in this case as demonstrated by Stroke et al.[15]

F. IMAGING THROUGH DISTURBING MEDIA

It is important to know how badly holographic image formation is affected by an inhomogeneous medium between the object of interest and the hologram recording medium. Such a medium distorts the waves on their way to and from the object. Such a medium images just like any other object and may block the view of the object of interest. Hopefully we can remove the effect of the intervening medium altogether. Barring that, we can try to eliminate the imaging of the medium so the object of interest can be seen.

Elimination of the effect of the intervening medium is accomplished in either of two ways, depending on whether the intervening medium is stationary or moving during the hologram exposure. If the medium is stationary, the intervening medium can be considered to be an aberration producer, and the aberration reduction technique just discussed (Section VIII-E) will remove the effect of the medium. There are many variations of this technique. Especially successful have been those due to Kogelnik,[16] Leith and Upatnieks,[17] and Kogelnik and Pennington.[18] These methods, like aberration correction, require a knowledge or record of the inhomogeneous medium. If the intervening medium moves while the object stands still, we can discriminate against the noise produced by the medium on that basis. Techniques of Spitz[19] and Stetson[20] accomplish this by utilizing the fringe degradation caused by moving objects (scatterers in this case).

Unfortunately, many real cases will preclude either of the two types of techniques used to eliminate the effect of the intervening medium, in which case we can at least keep the intervening scatterers from being imaged. This was accomplished by Caulfield[21] by adjusting the configuration in such a way that the object is in that portion of the coherence volume (Section VII-D) nearest the hologram. Thus, no intervening scatterers can be imaged.

A method which combines time averaging with deconvolution for the case in which the reference beam and object beam traverse essentially the same incoherent region was introduced by Goodman et al.[22] and analyzed in detail by Gaskill.[23] Under these circumstances, excellent images can be taken through shower glass or equivalent time-varying media.

G. MULTIPLE-IMAGE PHOTOGRAPHY

Integral photography is the name given to Lippman's[24] technique of multistereo photography. In integral photography, an array of lenses (a "fly's-eye lens") is used to form multiple, spatially distinct images of a scene on the back focal plane of the array. Each image is a complete image as seen by a particular lenslet in the array. Thus, the position on the photographic image records the direction of viewing and the image records the intensity. Both phase (direction) and intensity are recorded and can be reconstructed. In reconstruction a positive transparent image is repositioned at the focal plane and is illuminated by diffuse light At the image (where the object was) each lenslet image on the transparency gives the proper view of the object when viewed from the direction of that image. Thus, a three-dimensional image of the object is available. This three-dimensional image is limited in its "depth" by the number of lenslets in the array and limited in its resolution by the size of each lenslet. These opposing trends necessitate a compromise since the lenslets cannot overlap. The multiple-imaging technique developed

Figure VIII-2 Holograms need not be confined to small objects. These views of an outdoor scene are taken from an image produced by Sun Lu's multi-image hologram. From Lu, *Laser Focus*, 5 (3), 36 (1969).

by Lu[25] permits each lens to utilize the whole area, but involves compromising the energy gathered by each lens by a factor of $1/fN$.[26] Here N is the number of lenses and f is the packing fraction ($f < 1$) for ordinary fly's-eye lenses.

The great advantage of integral photography is that it does not involve interference. Therefore the light and stability requirements are the same as those for ordinary photography.

The great disadvantages of integral photography are threefold. First, it requires a careful reconstruction with a fly's-lens. Second, it produces the effect of viewing the image through a screen which corresponds to the unutilized regions of the fly's-eye lens (since nonoverlapping circles can not fill a square area). Third, the image is "pseudoscopic" (depth-inverted). In 1967 Pole[27] found a way to overcome these limitations. He reconstructed the image in laser light and recorded it as a hologram. The hologram was easily copied and reconstructed. Furthermore, it provided means for removing the "screen effect" at some cost in image resolution.[28] The pseudoscopy is removed by holography (which normally produces a pseudoscopic real image), since a doubly pseudoscopic image is orthoscopic.

McCrickerd and George[29] followed Pole's paper[27] with one discussing the replacement of the fly's-eye lens (both in the taking and reconstructing stages of integral photography) by one lens moved sequentially through an array of positions. This technique has been used by Lu to produce a hologram of an outdoor scene (a significant increase in subject size over the toy trains and chessmen usually used for holography) (see Figure VIII-2).

Several other novel applications of multiple-image holography have been described. These make multiple views of the object available by using multiple images (holographic or ordinary) superimposed on the same hologram.[30,31] Many more variations are obvious.

H. HOLOGRAPHY WITHOUT VISIBLE LIGHT

There are several reasons we might want to use something other than visible light for holography. First, we might be able to penetrate regions inaccessible to visible light. For example, X-rays and ultrasonic waves penetrate human flesh. Second, we might wish to utilize the power or coherence of some nonvisible-wave source, for example, a CO_2 laser for power or a piezoelectic transducer for coherence. Third, we might want to avoid heavy transmission losses for long distance holography. Sonic and ultrasonic waves penetrate water well and, hence, would be useful for underwater imaging. Fourth, holography may offer improved resolution with a given wave source. This was Gabor's motivation in inventing holography (for electron microscopy).

There are numerous problems which arise in holography without visible light. Most of these stem from the lack of suitable recording media.

There are also numerous advantages which nonvisible waves can offer over visible light for holography. Most of these occur because some nonvisible waves are slow varying so that their phases can be directly recorded.

The types of nonvisible-wave holography are so varied that we have chosen not to deal with them in detail here. Rather we cite representative papers on electron waves,[32,33] radiofrequency and microwaves,[34,35] sonic and ultrasonic waves,[36-42] ultraviolet light,[43] and infrared light.[44] Holography using the CO_2 laser and X-rays is discussed in Chapter XVI.

I. SPECIAL-PURPOSE HOLOGRAPHY

A number of special-purpose holographic systems have been devised. Some of these will be listed here.

Holographic portraiture has become possible as a result of work by Siebert.[45,4z] Beside the obvious commercial market, there are other application areas, such as prosthetics and dental plate fittings. Unfortunately, the initial investment cost precludes many of these applications at the moment, however, this will certainly change in time. Figure VIII-3 shows a photograph of the image reconstructed from one of Siebert's holograms.

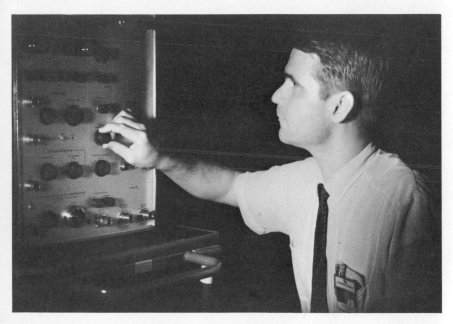

Figure VIII-3 L. D. Siebert[45,46] has pioneered the field of holographic portraiture. With his assistance and permission, we show a photograph of the image from one of his holograms. Siebert has paid careful attention to safety measures[46] and strongly recommends a diffuse light source to avoid eye damage.

Holograms can be used to generate contours of constant depth[47-49] or constant Doppler shift.[50]

Holograms for various display needs can be made.[51] In particular, panoramic displays (viewable over wide angles up to 360°) can be made in a number of ways.[52-54]

REFERENCES

1. E. N. Leith and J. Upatnieks, *J. Opt. Soc. Amer.*, **54**, 1295 (1964).

2. Y. N. Denisyuk, *Opt. Spectros.* **15**, 279 (1963).

3. G. Lippman, *J. Phys.*, **3**, 97 (1894).

4. K. S. Pennington and L. H. Lin, *Appl. Phys. Letters*, **7**, 56 (1967).

5. R. J. Collier and K. S. Pennington, *Appl. Opt.*, **6**, 1091 (1967).

6. L. H. Lin, K. S. Pennington, G. W. Stroke, and A. E. Labeyrie, *Bell Syst. Tech. J.*, **45**, 659 (1966).

7. A. A. Friesem and R. J. Fedorowiz, *Appl. Opt.*, **5**, 1085 (1966).

8. J. Upatnieks, J. Marks, and R. J. Fedorowiz, *Appl. Phys. Letters*, **8**, 286 (1966).

9. A. A. Friesem and R. J. Fedorowiz, *Appl. Opt.*, **6**, 529 (1967).

10. W. H. Carter, P. D. Engeling, and A. A. Dougal, *IEEE Trans. Quant. Electron.*, **2**, 44 (1966).

11. A. H. Lohmann, *Appl. Opt.*, **4**, 1667 (1965).

12. W. A. Shurcliff and S. S. Ballard, *Polarized Light*, Van Nostrand, Princeton, 1964.

13. M. E. Fourney, A. P. Waggoner, and K. V. Mate, *J. Opt. Soc. Amer.*, **58**, 701 (1968).

14. J. Upatnieks, A. Vander Lugt, and E. N. Leith, *Appl. Opt.*, **5**, 589 (1966).

15. G. W. Stroke, G. Indebetouw, and C. Puech, *Phys. Letters*, **26A**, 443 (1968).

16. H. Kogelnik, *Bell Syst. Tech. J.*, **44**, 2451 (1965).

17. E. N. Leith and J. Upatnieks, *J. Opt. Soc. Amer.*, **56**, 523 (1966).

18. H. Kogelnik and K. S. Pennington, *J. Opt. Soc. Amer.*, **58**, 273 (1968).

19. E. Spitz, *Compt. Rend.*, **264B**, 1449 (1967).

20. K. A. Stetson, *J. Opt. Soc. Amer.*, **57**, 1060 (1967).

21. H. J. Caulfield, *J. Opt. Soc. Amer.*, **58**, 276 (1968).

22. J. W. Goodman, W. H. Huntley, Jr., D. W. Jackson, and M. Lehmann, *Appl. Phys. Letters*, **8**, 311 (1966).

23. J. D. Gaskill, *J. Opt. Soc. Amer.*, **58**, 600 (1968).

24. G. Lippmann, *Compt. Rend.*, **146**, 446 (1908).

25. S. Lu, *Proc. IEEE* **56**, 116 (1968).

26. H. J. Caulfield, S. Lu, and J. L. Harris, *J. Opt. Soc. Amer.*, **58**, 1003 (1968).

27. R. V. Pole, *Appl. Phys. Letters*, **10**, 20 (1967).

28. C. B. Burckhardt, R. J. Collier, and E. T. Doherty, *Appl. Opt.*, **7**, 627 (1968).

29. J. T. McCrickerd and N. George, *Appl. Phys. Letters*, **12**, 10 (1968).

30. J. D. Redman, *J. Sci. Instr.*, **1**, 821 (1968).

31. J. D. Redman, W. P. Wolton, and E. Shuttleworth, *Nature*, **220**, 58 (1968).

32. M. E. Haine and T. Mulvey, *J. Opt. Soc. Amer.*, **42**, 763 (1952).

33. A. Tonomura, A. Fukuhara, H. Watanabe, and T. Komoda, *Japan. J. Appl. Phys.*, **7**, 295 (1968).

34. R. P. Dooley, *Proc. IEEE*, **53**, 1733 (1965).

35. G. Tricoles and E. L. Rope, *J. Opt. Soc. Amer.*, **57**, 97 (1967).

36. R. K. Mueller and N. K. Sheridan, *Appl. Phys. Letters*, **9**, 328 (1966).

37. K. Preston and J. L. Kreuzer, *Appl. Phys. Letters*, **10**, 150 (1967).

38. A. F. Metherell, H. M. A. El-Sum, J. J. Dreher, and L. Larmore, *Appl. Phys. Letters*, **10**, 277 (1967).

39. A. F. Metherell, H. M. A. El-Sum, and L. Larmore, Eds., *Acoustical Holography*, Plenum Press, New York, 1968.

40. G. A. Deschamps, *Proc. IEEE*, **55**, 570 (1967).

41. G. A. Massey, *Proc. IEEE*, **55**, 1115 (1967).

42. R. B. MacAnally, *Appl. Phys. Letters*, **11**, 266 (1967).

43. R. F. Wuerker, L. O. Heflinger, and R. A. Briones, *Appl. Phys. Letters*, **12**, 302 (1968).

44. R. F. van Ligten and K. C. Lawton, *J. Opt. Soc. Amer.*, **58**, 1556 A (1968).

45. L. D. Siebert, *Appl. Phys. Letters*, **11**, 326 (1967).

46. L. D. Siebert, *Proc. IEEE*, **56**, 1243 (1968).

47. K. Haines and B. P. Hildebrand, *Phys. Letters*, **19**, 10 (1965).

48. B. P. Hildebrand and K. Haines, *J. Opt. Soc. Amer.*, **57**, 155 (1967).

49. N. Shiotake, T. Tsuruta, Y. Itoh, J. Tsujiuchi, N. Takeya, and K. Matsuda, *Japan. J. Appl. Phys.*, **7**, 904 (1968).

50. J. W. Goodman, *Appl. Opt.*, **6**, 857 (1967).

51. B. P. Hildebrand, *Inform. Display*, **5**, (2), 28 (1968).

52. E. P. Supertzi and A. K. Rigler, *J. Opt. Soc. Amer.*, **56**, 524 (1966).

53. T. H. Jeong, R. Rudolph, and P. Luckett, *J. Opt. Soc. Amer.*, **56**, 1263 (1966).

54. Sh. D. Kakichashvili, A. I. Kovaleva, and V. A. Rukhadze, *Opt. Spectroses*, **24**, 333 (1968).

IX

Applications to 2-D Photography

A. INTRODUCTION

In comparison to conventional photography, holography is an alternative means to record an image. It has the unique advantage of recording phase, as well as intensity information of a wavefront and thus can reconstruct a three-dimensional image of the recorded object. In the first impression, holography does not offer any advantage over conventional photography in recording any object which has no third dimension. However, due to the unique way that holographic pictorial information is recorded, the potential application of holography in this field is well justified. For example, in most cases the pictorial information is uniformly recorded so that scratches, fingerprints, and dust particles on the hologram plane will not have direct image degradation, but only produce background noise. Also, a whole reconstructed image can be obtained from a small broken piece of a hologram. Holography can accurately reproduce the relative grey tone of a picture in a very wide range, while in conventional photography, the reproduction of highlights and shadow details of a scene is limited by the finite dynamic range of the recording medium. Furthermore, the relative grey tone of the scene is distorted by the inherent nonlinearity of photography (see Chapter V).

In many applications of forming extremely precise images, such as for integrated circuits masking, holography offers a much cheaper way to obtain a high resolution image; comparable conventional systems might be too expensive to build. It can hardly be said that the image formed by a hologram will have some better features than those formed by a lens, but holography has high potential as a cheaper and easier way to attain the ultimate image quality of a diffraction-limited system.

Another very important advantage of holography comes from the use of a reference beam in recording the image. A hologram can be recorded by using a coded reference beam in such a way that the recorded image can only be obtained by reconstructing the hologram with the proper beam of light. Leith and Upatnieks[1] have shown that a diffuser can be placed in the reference beam to record a hologram. The reconstructed image can only be obtained by aligning the diffuser and the hologram exactly in the same positions as they occupy in the recording process. The diffuser in the reference beam can be considered as a very complicated phase plate; it modulates the reference beam into a highly complicated wavefront and thus creates a code to carry the picture information onto the hologram. An obvious application associated with this feature is the recording of secret documents. Another very promising application is the use of reference waves of different spatial frequency to record many images on a single piece of recording medium. For example, by the use of thick emulsion, holograms can be made to have a diffraction efficiency very sensitive to the angle at which the reconstruction wave impinges on the hologram. Many images can be stored on the same hologram with reference waves of different spatial frequency. The images can be read out one by one as the angle of the reconstruction beam is changed. A very large amount of information can be stored on such a multiexposed hologram. Perhaps, someday, a whole book can be stored on a single hologram, and an individual page of the book can be projected on a screen by scanning a reconstruction beam (see Chapter XIV).

B. PROJECTION PRINTING

The recent development of integrated circuit technology creates a need to produce extremely sharp and precise images of circuit patterns with dimensions down to 2.5μ. In most cases a mask of the circuit pattern array is formed in a piece of photographic plate, and the image is transferred by contact printing to the semiconductor wafer, which is coated with photosensitive resist.

The developed photoresist layer performs the masking function in the subsequent diffusion or etching process. It is very essential (because of the complexity of modern integrated circuits) that each masking be almost perfect so that a high percentage yield may be obtained. In order to achieve the required image resolution, a very close contact between the mask and the wafer is necessary. The sizable contact pressure between the two reduces the useful life of a mask and sometimes also causes imperfections in the photoresist layer, thus affecting the neatness of the masking. In principle, the problems associated with contact printing can be removed by using image projection. Unfortunately, a conventional projection system with the required resolution and precision is very expensive and difficult to produce by present technology. In the search for an alternative which will achieve the required projection system, we turn to holography. For example, we can construct a hologram of a mask and use the real, reconstructed image to expose the photoresist layer on the semiconductor wafer. In this case no conventional imaging systems (lenses and mirrors) are required. The only cost will be for the photographic plate and the labor to record the hologram. Figure IX-1 shows the way to use a hologram for projection printing.

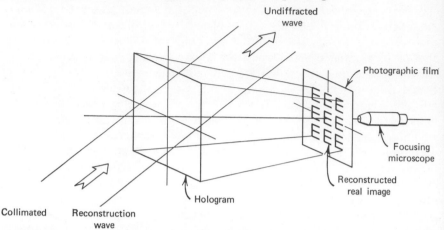

Figure IX-1   A schematic diagram showing how holography can be used for projection printing.

C. IMAGE QUALITY

It is worthwhile to compare the image resolution obtained from a holo- gram with those from a lens. A lens is usually specified by its f-number, which is the focal length of the lens divided by the diameter of the aperture. The smaller the f-number, the greater the light-gathering power. The f-number also determines the theoretical resolution limit of the lens set by the diffrac- tion of light. This limit is usually given in terms of the number of resolvable elements per unit length of the image field by the following equation:

$$\text{R.L.} = \begin{cases} \dfrac{1}{\lambda(f\text{-number})} & \text{for line images} \\[3mm] \dfrac{0.82}{\lambda(f\text{-number})} & \text{for point images,} \end{cases} \qquad \text{(IX-1)}$$

where λ is the wavelength of the light. Obviously, from this equation, a low f-number lens has a high theoretical resolution limit. In practical cases the theoretical limit cannot be achieved. Other factors, chiefly aberrations, put a limit on the maximum resolution long before the diffraction limit given in Eq. (IX-1) can be approached. Aberrations exist not only with lenses, but also with holograms, cathode ray tubes, and all other imaging devices. The major aberrations of an imaging system are the third-order aberrations which can be grouped into spherical aberration, coma, distortion, astigmatism, and curvature of field. Aberrations can be compensated for, to a certain extent, by using multi-element lens design; they can also be reduced by stopping down the aperture of the lens. The resolution of a lens can be reduced considerably by the existence of aberration; hence, by stopping down the aperture, a higher resolution can usually be achieved until finally the aperture of the lens become so small that the diffraction limit takes over and the resolution of the lens goes down again. A good example can be found with ordinary camera lenses. A typical 35 mm camera lens with f-number 2.0 has a resolution of about 40 lines/mm when the lens is wide open (at f-number 2.0). The resolution reaches about 80 lines/mm at medium f-numbers between 5.6 and 8.0 and finally drops off to 50 lines/mm at f-number 16.

A hologram has an effective focal length as well as an effective aperture. The effective focal length of a hologram is simply the distance between the hologram and the real image formed with a collimated reconstruction beam. It is easy to make a hologram with an effective f-number less than 1.0. Such holograms have a theoretical resolution limit of about 1600 lines/mm with 6328 Å light. The effective aperture of the hologram may not be equal to its physical aperture. This is caused by the resolution limit of the holographic recording medium as noted in Chapter V. In some holograms the spatial frequency of the diffraction pattern from the object becomes so high at the edges of the hologram that the recording medium ceases to record it; the outcome is a small, effective aperture less than the physical aperture of the hologram. Fortunately, with recording media such as Kodak 649F plates, the resolution limit, due to this effect, is better than 1μ. This is quite adequate for applications in projection printing.

The aberrations associated with a hologram have been investigated by many authors [2-4]. In theory, aberrations (at least the third-order aberrations) can be minimized in general cases. Readers who are interested in details may read any one of those three references. In experimental work with projection printing by holography, it is quite difficult to obtain image magnification without introducing aberrations. Fortunately, in the applications of projection printing to integrated circuit masking, it is desirable to work at unit image magnification.

In general, the ultimate image quality from a hologram depends on the use of a reconstruction wavefront which reproduces the reference wavefront as closely as possible. Any difference between the two will cause aberrations of the image. For this reason, the best image is formed by holograms with a plane reference wavefront and a plane reconstruction wavefront. The major aberration with the image is astigmatism, in which case the effect becomes dominant when the geometry of the image gradually approaches micron order. Even with plane reference wavefronts and plane reconstruction wavefronts, astigmatism can still be observed. It can be minimized by aligning the

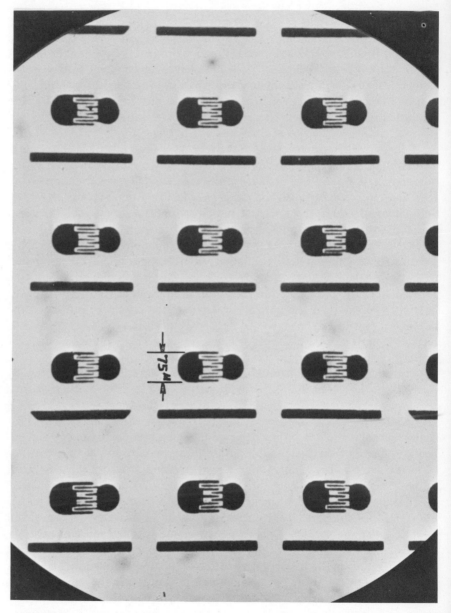

Figure IX-2 A magnified portion of an image array projected by a hologram. Courtesy of Texas Instruments Incorporated.

hologram with micro positioning devices to let the reconstruction beam reproduce the angle of incidence of the reference beam (an accuracy of several minutes is necessary).

Figure IX-2 shows a magnified portion of an image array projected by a hologram. Figure IX-3 shows the same image taken with 400 × magnification.

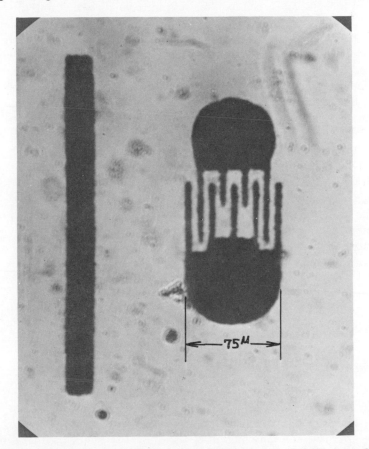

Figure IX-3 The same image as shown in Figure IX-2 but magnified 400 times. Courtesy of Texas Instruments Incorporated.

When all the aberrations have been removed, the image produced by a hologram still does not reach the ultimate quality of those from a well-corrected lens. The problem is associated with the inherent nature of coherent radiation. Imaging with coherent light creates a much noisier image than that obtained with incoherent light. In the coherent imaging system, any point that scatters light will cause a diffraction pattern superimposed on the image and ruin its clarity. On the other hand, with incoherent light, different frequency components produce different diffraction patterns, thus "averaging out" the noise. A lens is essentially designed such that it can be used with incoherent light. A hologram can produce a high resolution image only with coherent light. It is for this reason that the image formed by the best lens is still superior to a holographic image.

When holography is used for precision microimage projection, the use of diffused light to illuminate the object will severely degradate the image quality. A hologram taken with diffused light can record the picture information more uniformly, resulting in higher total reconstruction efficiency. However, the randomly scattered light from a diffuser causes speckling patterns on the image which not only break up the image into small specks, but also mar its fine details. Holograms taken without using diffused illumination have much clearer images; however, the speckling patterns are still not completely removed. In these cases it is the grain in the photographic emulsion which scatters light and causes a speckling pattern. If extremely fine grain film is used to record the hologram, the speckling pattern, produced by the photographic grain, only slightly increases the noise level of the reconstructed image.

If the projected image from a hologram involves fine details of very small dimensions, the shrinkage of the emulsion on the hologram after development can distort the pattern slightly. The effect is minor, but troublesome.

To obtain high quality, high resolution images with holography is a relatively easy process. It is difficult to maintain the same high quality and resolution over a large image field. To achieve this end, extremely careful control of the hologram recording parameters and the involved photographic process is necessary.

D. MULTIPLE-IMAGE PROJECTION

A very interesting technique of simultaneous multiple-image projection through the use of a Fourier transform hologram is proposed by Lu.[5] The method can project a large array of very high quality multiple images of a master pattern without any motion of the optical system. In the past, other methods of simultaneous multiple-image generation, based on classical optics, have been proposed (e.g., an array-of-pinholes camera and the fly's eye lens—an array of small lenslets). None of these can be considered as high quality imaging devices. The image resolution is usually low and aberrations are very severe.

The basic principle of projecting multiple images by the Fourier transform hologram is illustrated in Figure IX-4. When a function $f(x)$ forms convolution integration with an array of δ functions, a multiple array of $f(x)$ functions results. In optics the convolution integration of two functions is performed by the optical spatial filtering technique (Chapter XI). The concept of optical

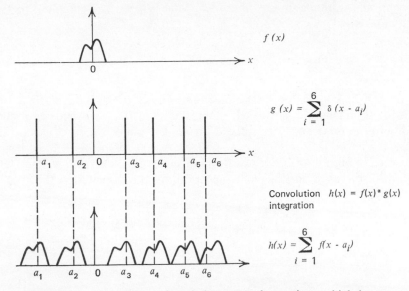

$$f(x)$$

$$g(x) = \sum_{i=1}^{6} \delta(x - a_i)$$

Convolution $h(x) = f(x) * g(x)$
integration

$$h(x) = \sum_{i=1}^{6} f(x - a_i)$$

Figure IX-4   A Fourier transform hologram used to project multiple images.

spatial filtering is based on the ability of a lens to perform Fourier transform between its focal planes with coherent light (Section II-B). It follows directly that if the optical disturbance passing through a transparency in one focal plane of the lens L_1 is given by a complex function $f(x', y')$ (Figure IX-5),

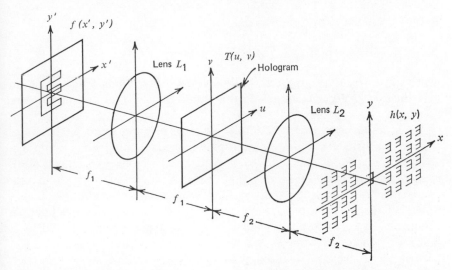

Figure IX-5   The Fourier transform relationship between focal-plane wavefronts using a lens in coherent light.

the optical disturbance in the other focal plane is given by a complex function $F(u, v)$ where

$$F(u, v) = \int\int_{-\infty}^{\infty} f(x', y')e^{-i(k/f_1)(ux' + vy')} \, dx' \, dy'. \tag{IX-2}$$

In Eq. (IX-2), $k = 2\pi/\lambda$ is the propagation constant of the coherent light, and f_1 is the focal length of lens L_1. If there is a mask with a complex amplitude transmitance function $T(u, v)$ placed on the u–v plane, the light leaving the mask is the product of $F(u, v)$ and $T(u, v)$. A second lens with focal length f_2 is used to form the Fourier transform of this product on the x–y plane. The optical disturbance on this plane, according to the convolution theorem, is given by

$$h(x, y) = \int\int_{-\infty}^{\infty} F(u, v)T(u, v)e^{-i(k/f_2)(ux + vy)} \, du \, dv$$

$$= \int\int_{-\infty}^{\infty} f(mx'', my'')t(x - x'', y - y'') \, dx'' \, dy'', \tag{IX-3}$$

where $t(x, y)$ is the Fourier transform of $T(u, v)$ and $m = -f_1/f_2$. For multiple-image generation through the optical convolution process, $t(x, y)$ must be an array of δ functions, that is,

$$t(x, y) = \sum_{p, q} \delta(x - a_p) \, \delta(y - b_q). \tag{IX-4}$$

Inserting this $t(x, y)$ into Eq. (IX-3), we obtain

$$h(x, y) = \sum_{p, q} f[m(x - a_p), m(y - b_q)], \tag{IX-5}$$

which physically is an array of multiple images. Here, m is a magnification or reduction factor; its negative sign simply means that the images are inverted.

In optics an array of δ functions is an array of point sources. Therefore, the key mask in u–v plane for multiple-image generation must contain a complex transmission function $T(u, v)$ which is the Fourier transform of an array of point sources. This is realized by using a Fourier transform hologram of the array of point sources. Since a hologram is used, the multiple images formed on the x–y plane include a bright, undiffracted image and two multiple arrays of images corresponding to the two conjugate images formed by a hologram. Figure IX-6 shows a 5 by 5 array of multiple images of a test pattern. The images are produced by using a Fourier transform hologram of a 5 by 5 array of point sources.

In this multiple-image generation system, for any master pattern put on the input plane, an array of multiple images will appear on the output plane. The Fourier transform hologram on the u–v plane can be considered as a multiple beam splitter, and the whole system is a telescopic imaging device with a hologram to provide the multiple beam splitting and, thus, multiple-image projection.

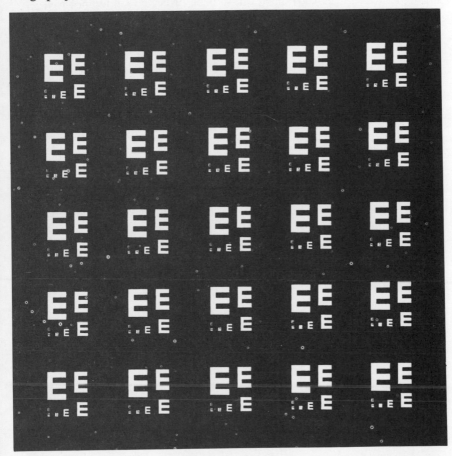

Figure IX-6 5×5 array of images projected by the method shown in Figure IX-4.

The advantage of this method over the classical methods is its capability of projecting images of very high quality. For example, with a pair of well corrected lenses, multiple images have been projected over a 5-cm field with 3μ resolution. The aberrations are unnoticeable, and image contrast is excellent.

The total number of multiple images projected is equal to the total number of point sources recorded on the hologram. The Fourier transform hologram of the array of point sources is constructed by using the multiple-exposure technique (Section VIII-G). It is found that the reconstruction efficiency of a hologram with a large number of multiple exposures is quite low. This is the chief limitation of projecting a large number of multiple images. However, in one experiment, over 1000 images have been projected simultaneously.

E. POWER APPLICATIONS OF HOLOGRAPHY

One of the most important uses of a laser is as a source of power for welding, drilling, surgery, etc. Many of these applications are handicapped by undesirable features of the focused laser beam. Many more applications would be opened if the shape of the concentrated laser beam could be predetermined. A uniform focal spot would be much more useful for some applications than the nonuniform focused beam produced by a lens. A square cross section beam may be more useful than a circular one. The possibilities are great. The use of highly transparent hologram media or reflective media and insertion of the hologram into the unfocussed beam should suffice to minimize damage to the hologram. The total efficiency of the system could reach 25 to 30%, a tolerable situation considering the immense powers now available from lasers.

REFERENCES

1. E. M. Leith and J. Upatnieks, *J. Opt. Soc. Amer.*, **56**, 523 (1966).
2. J. A. Armstrong, *IBM Technol.*, **9**, 171 (1965).
3. R. W. Meier, *J. Opt. Soc. Amer.*, **55**, 987 (1965).
4. G. L. Rogers, *Proc. Roy. Soc. (Edinburgh)*, **A63**, 313 (1952).
5. S. Lu, *Proc. IEEE*, **56**, 116 (1968).

X

Applications of Computer-Generated Holograms

A. INTRODUCTION

We now examine the problem of making a hologram given only the mathematical description of the object. For all but the simplest objects, the calculations involved become quite involved and digital computers must be used. The various techniques described below are well established and straightforward.

Several problems arise in all of these techniques, so these will be discussed first.

The various techniques all produce a mathematical description of a transparency which constitutes the hologram. Unfortunately, there have been no techniques devised which can produce that transparency directly because of the high resolution required. Therefore, the hologram must be plotted on an expanded scale and then photographically reduced. A wide selection of mechanical and electronic plotters exists, and the final result is independent of the plotter used.

All the techniques produce sampled holograms. The problem of "optimum" sampling for holography is just being formulated (see Section XVI-E), so great changes in this area may be forthcoming.

In all cases we must calculate a diffraction pattern. In all but the simplest cases, one technique is so much superior to the others in terms of computer time. This is the Cooley-Tukey[1] algorithm for calculation of digital Fourier transforms. This technique is sometimes called the "fast Fourier transform technique" because it requires only $N \log_2 N$ operations to compute N points on a diffraction pattern as opposed to N^2 operations using brute force methods. For 10^6 elements this is 2×10^7 for the Cooley-Tukey algortihm and 10^{12} for the brute force method. This difference makes computer-generated holograms practical. Indeed, the Cooley-Tukey algorithm can be utilized to calculate the sampled hologram directly,[2] but other approaches are easier and more practical.

B. COMPUTER-GENERATED SPATIAL FILTERS

The first computer-generated hologram known was not used as a hologram, but as a complex spatial filter. To do this Kozma and Kelley[3] calculated the continuous hologram and then sampled it. As noted above, this technique is impractical for many cases. The idea of computer-generated spatial filters was recently revived and improved by Lohmann and Paris,[4] who used the binary mask technique which we will now describe.

C. BINARY HOLOGRAMS

Perhaps the most revolutionary technique for computer-generated holograms is the binary mask method of Brown and Lohmann.[5] Rather than simply copying the optical hologram, they started over and asked, "How can we encode the amplitude and phase on a computer-generated mask in such a way that they are properly reconstructed when illuminated by a collimated beam of light?" Their answer was very simple. The amplitude can be controlled by the size of the aperture (so long as the "image" of that aperture is larger than the image of the "object" in the Fourier transform plane). The relative phase between two points can be controlled by the spacing (grating constant) between them. Thus, Brown and Lohmann divided the hologram plane into numerous square cells. Within each cell they placed a rectangular slit. The area of the slit was proportional to the amplitude calculated for the Fourier transform of the object at the center point of the cell. The displacement of the slit from the center of the cell was governed by the phase calculated for the Fourier transform of the object at the center of the cell. Further details on this method can be found in a paper by De and Lohmann.[6]

D. ZONE PLATE HOLOGRAMS

The construction of holograms by superimposed Fresnel zone plates has been a popular approach.[7,8] The analytical version of this technique is due to Waters.[9] Waters calculated the appropriate superposition of zone plates by computer. He then plotted those points having positive phase encountered when he sampled at an appropriate rate. His holograms also produced high quality images.

E. STRAIGHTFORWARD CALCULATION OF THE INTERFERENCE PATTERN

The simplest approach, conceptually, is to calculate the intensity of an optically generated hologram. This approach was first taken for very simple objects by Carter and Dougal[10] and later by Huang and Prasada.[2] The most successful of these techniques is that of Lesem et al.[11] Their technique permits the use of the Cooley-Tukey algorithm for calculating Fresnel diffraction patterns. The use of Fresnel diffraction allows unrestricted depth of field. The calculation technique of Lesem et al. can be combined with binary encoding if desired. All these techniques can be used to produce holograms of diffusely illuminated " objects."

F. COMPUTER-CONTROLLED HOLOGRAPHIC SYNTHESIS OF HOLOGRAMS

A point-by-point synthesis of holograms can be accomplished either by successive photographic superposition of zone plates[7,8] or by successive recording of a point source as it is moved through space.[13] Two photographs of such a hologram are shown in Figure X-1. In this case the number of

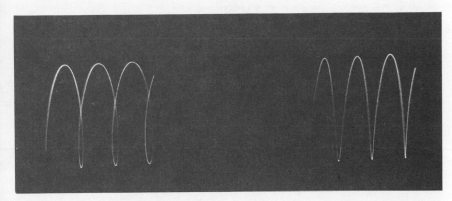

Figure X-1   Two views of the image, reconstructed from a hologram, made by more than 4000 consecutive exposures to a moving point source. From Caulfield, Lu, and Harris, *J. Opt. Soc. Am.*, **58**, 1003 (1968).

points in the image, N, is limited by the maximum signal-to-noise ratio, S_1, that can be achieved in a hologram of a single point and by the minimum signal-to-noise ratio, S_N, which can be tolerated for a single point. The relationship is

$$N = \left(\frac{S_1}{S_N}\right)^{1/2}.$$

The image shown in Figure X-1 contains about $N = 4000$ points.[13] Clearly this technique is not limited to point objects, so a library of movable elements (points, lines, spheres, etc.) could be used to synthesize composite three-dimensional images.

A further generalization of this technique was described by Lu et al.[14] They recorded the time-integrated interference pattern between light from a moving point source and light from a stationary point source to create a hologram of the orbit of the moving source. Thus, they were able to draw a picture in three dimensions on a two-dimensional photographic plate!

REFERENCES

1. J. W. Cooley and J. W. Tukey, *Math. Comput.*, **19**, 297 (1965).
2. T. S. Huang and B. Prasada, *Quart. Progr. Rept., Res. Lab. Electron, MIT*, **81**, 199 (1966).
3. A. Kozma and D. L. Kelley, *Appl. Opt.*, **4**, 387 (1965).
4. A. W. Lohmann and D. P. Paris, *Appl. Opt.*, **7**, 651 (1968).
5. B. R. Brown and A. W. Lohmann, *Appl. Opt.*, **5**, 967 (1966).
6. M. De and A. W. Lohmann, *Appl. Opt.*, **6**, 2171 (1967).
7. N. O. Young, *Sky Telescope*, **25**, 8 (1963).
8. W. J. Siemens-Wapniarski and M. P. Givens, *Appl. Opt.*, **7**, 535 (1968).
9. J. P. Waters, *Appl. Phys. Letters*, **9**, 405 (1966).
10. W. H. Carter and A. A. Dougal, *J. Opt. Soc. Amer.*, **56**, 1754 (1966).
11. L. B. Lesem, P. M. Hirsch, and J. A. Jordan, *J. Opt. Soc. Amer.*, **58**, 729 (1968).
12. G. W. Stroke, F. H. Westervelt, and R. G. Zech, *Proc. IEEE*, **59**, 109 (1967).
13. H. J. Caulfield, S. Lu, and J. L. Harris, *J. Opt. Soc. Amer.*, **58**, 1003 (1968).
14. S. Lu, H. W. Hemstreet, Jr., and H. J. Caulfield, *Phys. Letters*, **25A**, 294 (1967).

XI

Applications to Optical Data-Processing

A. INTRODUCTION

In recent years the technique of optical data processing has been receiving increased attention in the fields of science and technology. This increased popularity has been closely associated with the invention of the laser as a strong coherent source of illumination. Although many operations of optical data processing could be performed with incoherent illumination and the basic theory of the technique was formulated years before the invention of the laser, the full potential could not be reached without lasers and holography. Thus, in this chapter we are going to restrict our attention mainly to the techniques of optical data processing with coherent illumination.[1-4]

The input data for the optical processor is a spatial distribution of phase and/or intensity. Examples include geophysical array data, radar data, and ordinary optical images. These data are usually static and, hence, are recorded on photographic emulsions. No fully satisfactory real-time input has been invented as yet.

The main advantage of optical data processing is the speed of handling for extremely large amounts of data. This high speed is achieved because optical data is processed in a multichannel fashion. In fact, each resolution element on the input film can be considered a channel. With this assumption, it can be seen that it is not unusual to have many millions of channels processed at the same time.

Many of the optical data processing techniques are based on the concept of optical spatial filtering, in which the spatial frequency spectrum of the input optical imagery is modified by a spatial filter to achieve a predetermined purpose (e.g., edge sharpening, distortion removal) or to perform certain types of mathematical operations. In many applications of optical spatial filtering, the required filter function has to be complex (in the mathematical sense of having real and imaginary parts). That is, the spatial frequency spectrum of the input imagery must be modified both in amplitude and phase in order to achieve the predetermined purpose. In these cases holography is the only technique that allows the proper complex filter to be synthesized.

In the following, we are going to describe, first, the basic principles and techniques of optical spatial filtering and, then, the application of holography to synthesize the required filters for optical data processing.

B. OPTICAL SPATIAL FILTERING

The whole development of the optical spatial filtering techniques is based on the Fourier transform property of a lens. Through this property, the two-dimensional spatial frequency spectrum of a set of optical data can be displayed on a physical plane for filtering. There are several ways in which a positive lens can be used to obtain the two-dimensional Fourier transformation between a wavefront in plane P_1 and a wavefront in plane P_2.[5] An example is shown in Figure XI-1. In this setup both of the planes are located at

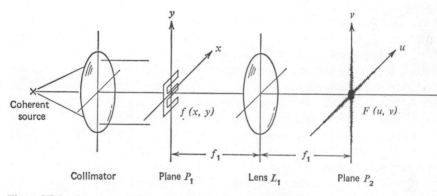

Figure XI-1   One way of using a lens to obtain a two-dimensional Fourier transformation, $F(u, v)$, of the complex function, $f(x, y)$.

focal distance f_1 from lens L_1. The optical data to be transformed are recorded on a transparency which is placed in plane P_1. A coherent plane wave is used to illuminate the transparency. If the field transmission function of the transparency is $f(x, y)$, it can be proved that the optical disturbance obtained on plane P_2 is:

$$F(u, v) = \int\int_{-a}^{a} f(x, y)e^{-ik(ux + vy)/f_1} \, dx \, dy$$

$$= \int\int_{-a}^{a} f(x, y)e^{-i(\omega_x x + \omega_y y)} \, dx \, dy, \qquad (XI-1)$$

in which u and v are the linear coordinates on plane P_2; $k = 2\pi/\lambda$; $\omega_x = ku/f_1$; $\omega_y = kv/f_1$; and a is the radius of L_1. In fact, a constant multiplier was set equal to one in Eq. XI-1, since that constant is not important in our discussion. It is recognized from Eq. XI-1 that the field displayed on plane P_2 is the finite $(a \neq \infty)$, two-dimensional Fourier transformation of the function $f(x, y)$. Physically, this is the frequency spectrum of the input data. Since the transformation is carried out from a set of spatial coordinates x and y to a set of variables ω_x and ω_y, whose dimensions are radians per unit length, the two-dimensional spectrum, obtained on plane P_2 is called a "spatial frequency" spectrum.

The derivation of the Fourier transform property of a lens is given by several authors [5-7] and is sketched in Section II-B. It is also noted that the $F(u, v)$ function given in Eq. XI-1 is only a very close approximation of the actual field displayed on plane P_2. Points not on P_2 are not related linearly to $f(x, y)$. To the extent that the tangent of any ray angle is equal to the angle, the approximation holds.

The importance of the Fourier transform property of a lens is not so much that it can display the spatial frequency spectrum of an input optical data on a plane, but that it makes it possible to modify both the amplitude and the phase of the spectrum simply by placing a filter on the spatial frequency plane, P_2. This arrangement forms the basis of the widely used technique of optical spatial filtering.

The spatial frequency spectrum of the input data, after being modified by the filter, is Fourier transformed by a second lens, L_2, of focal length f_2, and radius a. The resultant optical field appears on plane P_3 and is ready for readout. The whole setup for optical spatial filtering is shown in Figure XI-2.

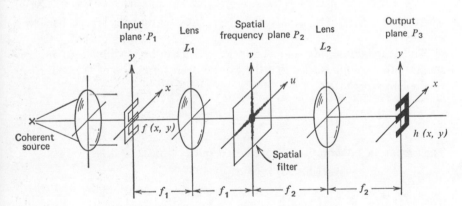

Figure XI-2 A generalized optical spatial filtering setup. Lens, L_1, is used to Fourier transform the input wavefront $f(x, y)$ in plane, P_1, to form its Fourier transform $F(u, v)$ in plane, P_2. A spatial filter is inserted at P_2. The filtered wavefront is retransformed by lens, L_2, to form an output distribution $h(x, y)$ in plane, P_3. The relationship of $f(x, y)$ to $h(x, y)$ is governed by the nature of the filter.

Let us find the mathematical expression for the optical field on the output plane, P_3. Suppose that the field transmission function of the spatial filter is given by $T(u, v)$, then the optical disturbance passing through the filter is $F(u, v)T(u, v)$. The optical field on the plane P_3 is therefore the Fourier transformation of this product, that is,

$$h(x, y) = \int\int_{-a}^{a} F(u, v)T(u, v)e^{-ik(xu+yv)/f_2}\, du\, dv. \qquad (XI\text{-}2)$$

From the defining equations, we know that the only difference between the inverse Fourier transformation and the direct Fourier transformation is the sign in the exponential kernel. In the inverse transformation, this sign is positive, and in the direct transformation, the sign is negative, as those in Eq. XI-2. However, if we invert the coordinate system by letting $x' = -x$ and $y' = -y$, Eq. XI-2 becomes an inverse Fourier transformation from the u and v coordinates to the new x' and y' coordinates. Therefore, in this inverted coordinate system, the function of lens L_2 is to transform the modified frequency spectrum back into an optical image.

It is evident from the above argument that if the filter is removed $[T(u, v) = 1]$, the unmodified frequency spectrum is transformed back and an image of the input optical data will be formed on the output plane. However, this image will be inverted as compared to the original input. Also, it produces a reduction (or magnification) if f_1 and f_2 are not equal. This reduction (or magnification) factor can be found from a simple geometric optics argument to be f_1/f_2. Therefore,

$$h(x, y) = f(mx, my) \qquad \text{if } T(u, v) = 1, \qquad (XI\text{-}3)$$

where $m = -f_1/f_2$. The negative sign reminds us that the image has been inverted.

The expression of $h(x, y)$ given in Eq. XI-2 can be modified into another standard form. Suppose $t(x, y)$ is the Fourier transformation of $T(u, v)$ formed by the lens L_2, that is,

$$t(x, y) = \int\int_{-a}^{a} T(u, v)e^{-i(k/f_2)(xu+yv)}\, du\, dv. \qquad (XI\text{-}4)$$

Then from the convolution theorem of Fourier transformation

$$h(x, y) = f(mx, my) \circledast t(x, y)$$

$$= \int\int_{-a}^{a} f(mx'', my'')t(x - x'', y - y'')\, dx''\, dy''$$

$$= \int\int_{-a}^{a} f[m(x - x''), m(y - y'')]t(x'', y'')\, dx''\, dy'', \qquad (XI\text{-}5)$$

where the notation "⊛" means the convolution integrations given in Eq. XI-5.

There are many operations of optical data processing that can be performed by this simple technique of optical spatial filtering. The following are examples that have been realized by experiments.

1. Image contrast improvement. The field transmission function of a low contrast picture has a large dc component. Hence, the image contrast can be increased by partially or completely removing the zero frequency component in the spectrum. The filter might be simply a transparency with a small attenuating spot at the origin of the frequency plane.

Another simple operation that uses a similar filter is the edge enhancement. The filter attenuates the low frequency components of the spectrum. The resultant image thus has sharp rising edges and is therefore easier to recognize.

2. Signal enhancement. A simple filter can be used to remove the pattern of equally spaced lines in a picture obtained by a scan technique. The scan lines produce a spectrum containing discrete components at the fundamental frequency and its harmonics. The superimposed picture information and scan line information in the input plane are convolved in the frequency domain, so the picture information can be considered as a modulation of the intensity at each of the discrete scan-line frequency components. A spatial filter in the frequency plane which blocks all but one of these components removes the scan-line information, but not the picture information. The resulting output image is therefore free from the scan lines (see Figure XI-3).

From a similar argument, a filter which is only transparent at a set of equally spaced points can be used to separate a periodic signal from a random noise.

3. Optical image improvement. Filters that only modify the amplitude of the spectrum can be used for apodization.[8,9] In these applications the spread function of an imaging system is sharpened by a properly constructed amplitude filter. The resolving power of the imaging system can therefore be increased.

Simple amplitude and phase filters have been used in correcting aberrant images.[10] The amount of wave aberration must be known. The filter is constructed to compensate the aberration. The input image can be improved to a certain extent.

4. Mathematical operations. By using a proper filter function on the spatial frequency plane, several mathematical operations can be performed on the input data. It is obvious that convolution integration between two functions is obtained by optical spatial filtering. The filter function that is necessary for these applications is usually a complex function.

Another example of a mathematical operation that can be performed by a spatial filter is differentiation. A one-dimensional differentiation of the input image requires a filter which has a field transmission function

$$T(u, v) = u \tag{XI-6}$$

It is quite simple to prove the above statement. Since $f(x, y)$ is the inverse Fourier transformation of $F(u, v)$, we have

$$f(x, y) = \int\limits_{-a}^{a}\!\!\int F(u, v)e^{ik(ux+vy)/f_1} \, du \, dv$$

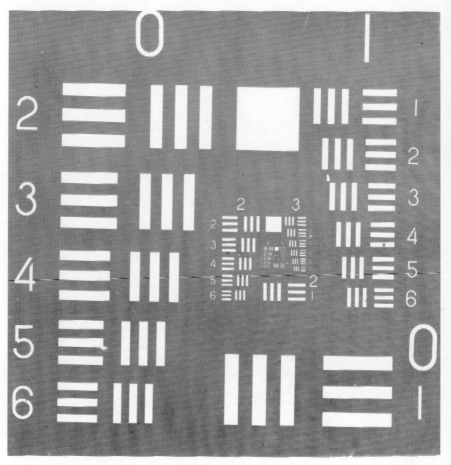

Figure XI-3 These photos include (a): the image with scan lines.

Figure XI-3b Its spectrum.

and

$$\frac{\partial}{\partial x} f(x, y) = \iint\limits_{-a}^{a} F(u, v) \frac{\partial}{\partial x} \left[e^{ik(ux + vy)/f_1} \right] du\ dv$$

$$= i\frac{k}{f_1} \iint\limits_{-a}^{a} uF(u, v)e^{ik(ux + vy)/f_1} du\ dv.$$

That is, the Fourier transformation of $\partial[f(x, y)]/\partial x$ is $F(u, v)u$. Thus, the spectrum of $\partial[f(x, y)]/\partial x$ can be produced on the spatial frequency plane by passing the spectrum of $f(x, y)$ through a filter $T(u, v) = u$.

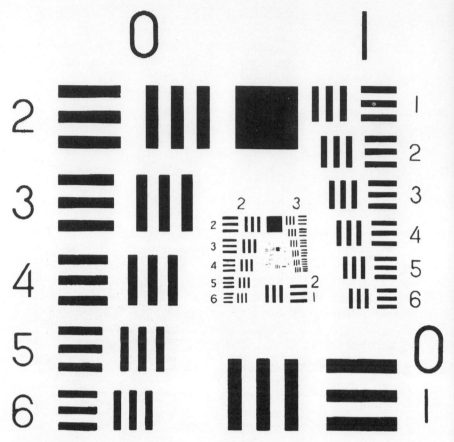

Figure XI-3c   The image after the scan lines have been filtered out. The spectrum includes components equally spaced along the vertical (wy) axis. The filter is constructed in such a way that only the component located around $wy = 0$ is passed. The filtered image subjects some loss of image quality for those horizontal line elements whose spatial frequency is close to or higher than those of the scan lines. For example, all the horizontal lines in groups 4 and 5 are not well reproduced. Even the horizontal elements in group 3 are affected.

Several other mathematical operations can be derived from Eq. XI-5. For example, the correlation function between two functions $f(x)$ and $t(x)$,

$$\phi_{ft}(x) = \int_{-\infty}^{\infty} f(x'')t(x + x'') \, dx'', \tag{XI-7}$$

is the same as the convolution integration between $f(-x)$ and $t(x)$, hence, the correlation function can be obtained through optical spatial filtering. Also, there are many integral transforms that can be reduced into a convolution integration. In principle, these transforms can be performed by using a proper spatial filter.

5. *Pattern and character recognition.* The use of an optical spatial filter to recognize a known pattern or character was demonstrated by Vander Lugt.[11,12] This application is based on the theory of matched filters.[13] According to the theory, if we want to detect a known signal $s(x, y)$ in a random, stationary, additive noise $n(x, y)$, the optimum filter (for maximum signal to noise energy ratio) is given by

$$T(u, v) = \frac{cS^*(u, v)}{N(u, v)}, \qquad (XI-8)$$

where $S(u, v)$ is the Fourier transform of $s(x, y)$, $N(u, v)$ is the noise power spectrum which is the Fourier transform of the autocorrelation function of the noise $n(x, y)$, c is a constant, and the notation * denotes the complex conjugate.

The input data $f(x, y)$ is given by $s(x, y) + n(x, y)$. It is noted that the matched filter only detects the existence of a known signal at a certain location. Since detection is contemplated by measurement at one point in space, the filter concentrates the signal enegy. In fact, the output represents the probability that the known signal exists at any particular point.† At a location where this probability is high, a bright light spot appears (see Figure XI-4).

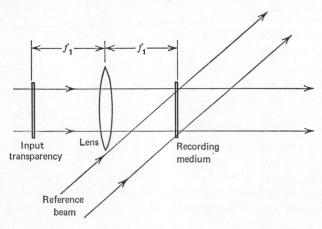

Figure XI-4 One method used to record the optical spatial filter by Fourier transform holography.

---

† The peak of the light spot is proportional to the cross correlation between the input character and the character to which the filter is matched. Identification is based on the fundamental theorem that the autocorrelation is at least as great as any cross correlation.

In the experimental work demonstrated by Vander Lugt[11,12] the known signal is either a pictorial pattern or a printed word of several alphabetical characters. The input data is a transparency of a printed page that contains the signal at some locations. The matched filter detects the locations of all the signals simultaneously, regardless of where the signals are.

Many other operations of optical data processing can be performed by optical spatial filtering or the techniques derived from it. In this book there are two more examples described in other chapters: (1) the use of spatial filters to correct certain lens aberrations (Chapter XIII) and (2) the use of spatial filters to generate multiple images from a single master input (Chapter IX).

C. APPLICATION OF HOLOGRAPHY IN OPTICAL SPATIAL FILTERING

It is understood from the above discussion that many useful operations of optical data processing require the use of spatial filters which have complex field transmission functions. These complex spatial filters can be satisfactorily realized only by using the technique of holography. In fact, any complex spatial filter is, by definition, a hologram.

In cases where the filter function is real and positive, the problem of synthesis becomes quite easy. A photographic plate can be selectively exposed to give the required filter function. However, an index matching liquid gate should be used with the filter in order to compensate for the unwanted phase shift due to the surface relief of the developed emulsion.

Very simple amplitude and phase filters have been synthesized by using an amplitude filter with a pure phase plate superimposed.[10] Spatial filters of real valued transmission functions (the phase is either 0 or 180°) have also been synthesized on Polaroid Vectograph film.[14]

In the case of matched filters, the filter function is given in Eq. XI-8. The denominator, $N(u, v)$, which is the power spectrum of the noise, is real and positive and, hence, can be realized physically without difficulty. In the case where there is only white noise in the input data (this is very unlikely to be the actual case), $N(u, v)$ is a constant, and the problem becomes even simpler. The numerator is, however, a complicated, complex-valued function. Therefore, the matched filter is very difficult to realize by the technique of amplitude and phase plates.

The first success of realizing a general complex spatial filter function by a simple process was that of Vander Lugt.[11] In his experiment an interferometric technique was used to record the complex filter function on a piece of photographic film. Later, this technique was found to be the same as that of Fourier transform holography.

Let us illustrate how a matched filter can be realized by a Fourier transform hologram. Figure XI-5 shows one of the many ways to record a Fourier transform hologram of a known signal, $s(x, y)$.* Obviously, lens L forms the Fourier transformation of the optical disturbance leaving plane P_1.

The known signal, $s(x, y)$, is placed on plane P_1. A point reference source formed by a focusing lens, L' is also introduced. Therefore the optical disturbance leaving P_1 is given by

$$g(x, y) = A\delta(x, y - d) + s(x, y), \qquad \text{(XI-9)}$$

where the point reference is assumed to be at $(0, d)$.

The optical disturbance on plane P_2 is

$$G(u, v) = \int\limits_{-a}^{a}\!\!\int [A\delta(x, y - d) + s(x, y)]e^{-ik(ux+vy)/f_1}\, dx\, dy$$

$$= Ae^{-ikdv/f_1} + s(u, v). \qquad \text{(XI-10)}$$

The first term in Eq. XI-10 is a plane reference wave, and the second term is the Fourier transform of the signal to be recorded

It is obvious by following the analysis given in Chapters III and V that the hologram has a field transmission function which is

$$T(u, v) = T_0 + c_1[e^{ikdv/f_1}s(u, v) + e^{-ikdv/f_1}s^*(u, v)]$$

$$+ \text{ higher order terms.} \qquad \text{(XI-11)}$$

The third term in Eq. XI-11 is the required complex function that forms the essential part of the matched filter. The phase factor e^{-ikdv/f_1} is the spatial carrier that makes the separation between the reconstructed wavefronts possible.

When such a hologram is used as a matched filter, the dc term forms a bright image of the unfiltered input data on the axis of the transforming lens. The filtered output appears at the location where the normal reconstructed image due to $S^*(u, v)$ appears. The other output due to the second term in Eq. XI-11 can be discarded. In Chapter IX we discussed a similar filter used for multiple-image generation. This basic technique is used for character recognition.

An alternative method of synthesizing a complex filter function was developed by Lohmann and Paris[15] (see Chapter X). Their idea was to use a digital computer to calculate the hologram that is used to realize the complex filter function. The details of these works are discussed in Chapter X and will

---

* The setup shown in Figure XI-5 is different from the original setup used by Vander Lugt. However, the principle involved is the same.

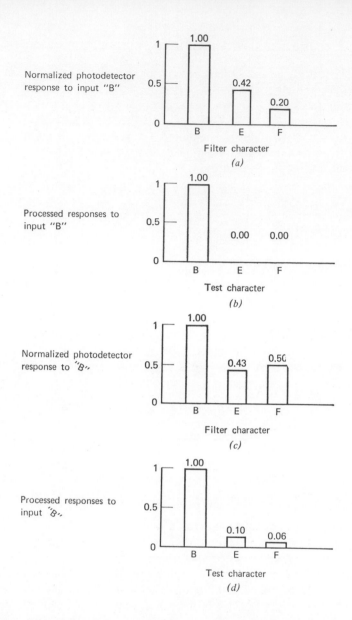

Figure XI-5 Optical spatial filters were made to identify the letters " B," " E," and " F." The response of these three filters to an input " B " is shown in (*a*). Linear combinations of those filters were chosen to set the cross-correlation terms to zero. Thus testing the input " B " for its identity to " E " or " F " yields a null response as shown in (*b*), where the test for " B " yields unit response. A rotated input " B " gave the responses shown in (*c*) to the original filters and the responses shown in (*d*) for the linear combinations of filters. For ease of comparison, the responses in each case (*a*, *b*, *c*, and *d*) were divided by the appropriate constant to set the " B–B " responses equal to one. Clearly, discrimination among characters has been enhanced by the linear combination technique.

106

not be repeated here. However, this method has been proved to be very useful.

Let us suppose that we wish to identify each input character as one of a finite set of N characters. The simplest method is to form all of the matched filters on a single photographic plate by multiple-exposure holography in such a way that the cross correlation spots fall in separate places in the image plane. These spots are read by N photodetectors which then produce, for input "character k," the output voltages $V_1(k)$, $V_2(k)$, ..., $V_N(k)$. The greatest of these voltages may be $V_n(k)$. We then say that the input character is identified as "character n." Hopefully, $n = k$.

A better way to utilize the output voltages is to form linear sums of the form

$$U_j(k) = \sum_{m=1}^{N} w_{jm} V_m(k).$$

We set the scale of the weights, that is, the $W|_{jm}$'s, by requiring $W_{jj} \triangleq 1$. We set cross correlations between characters to zero by requiring

$$U_j(k) = 0 \qquad \text{for } j \neq k.$$

Thus, for each j we solve N linear equations ($w_{jj} = 1$, $U_j(k) = 0$ for $k \neq j$, with N unknowns (the w_{jm}'s). These new outputs not only have zero cross correlation for ideal characters, but also show much improved tolerance to distorted inputs as noted in Figure XI-5 from the paper by Caulfield and Maloney.[16]

REFERENCES

1. E. L. O'Neill, *IRE Trans. Inform. Theory*, **2**, 56 (1956).

2. L. J. Cutrona, in *Optical and Electro-Optical Information Processing*, J. T. Tippett, D. A. Berkowitz, L. C. Clapp, C. J. Koester, and A. Vanderbugh, Jr., Eds., MIT Press, Cambridge, 1965, Chapter 6.

3. E. N. Leith, A. Kozma, and J. Upatnieks, in *Optical and Electro-Optical Information Processing*, J. T. Tippett, D. A. Berkowitz, L. C. Clapp, C. J. Koester, and A. Vanderbugh, Jr., Eds., MIT Press, Cambridge, 1965, Chapter 8.

4. A. Vander Lugt, *Opt. Acta*, **15**, 1 (1968).

5. J. W. Goodman, *Introduction to Fourier Optics* McGraw-Hill, New York, 1968.

6. L. J. Cutrona, E. N. Leith, C. J. Palermo, and L. J. Procello, *IRE Trans. Inform. Theory* **6**, 398 (1960).

7. E. B. Champagne, *Appl. Optics*, **5**, 1088 (1966).

8. G. Lansraux, in *Optical and Electro-Optical Information Processing*, J. T. Tippett, D. A. Berkowitz, L. C. Clapp, C. J. Koester, and A. Vanderbugh, Jr., Eds., MIT Press, Cambridge, 1965, Chapter 5.

9. P. Jacquinot and B. Roizen-Dossier, *Progr. Optics*, **3**, 31 (1964).

10. J. Tsujiuchi, *Progr. Opt.* **2**, 133 (1963).

11. A. Vander Lugt, *IEEE Trans. Inform. Theory*, **10**, 139 (1964).

12. A. Vander Lugt, in *Optical and Electro-Optical Information Processing*, J. T. Tippett, D. A. Berkowitz, L. C. Clapp, C. J. Koester, and A. Vanderburg, Jr., Eds., MIT Press Cambridge, 1965, Chapter 7.

13. G. L. Turin, *IRE Trans. Inform. Theory* **6**, 311 (1960).

14. T. M. Holladay and J. D. Gallatin, *J. Opt. Soc. Amer.*, **56**, 869 (1966).

15. A. W. Lohmann and D. P. Paris, *Appl. Opt.*, **7**, 651 (1968).

16. H. J. Caulfield and W. T. Maloney, *Appl. Opt.* **8**, 2354 (1969).

XII

Applications to Interferometry

A. BASIC THEORY

Interferometric holography offers a beautiful example of two aspects of holography. First, holography changes all the ground rules. Many obvious truisms of the past have proved false. There was a time when we believed that two wavefronts separated in either time or space could not interfere with each other and that two wavefronts of differing wavelength could not interfere with each other. But with holographic interferometry, these impossible events are the keys to startling new applications. Second, the growth of this field is so rapid that communication among holographers is not fast enough to prevent simultaneous discoveries. In the case of holographic interferometry, at least six groups discovered and published one or more variations of this idea at about the same time.[1-6]

The classical two-beam interferometer has a reference beam, an object, and an observation plane. The interference pattern at the observation plane is interpreted in terms of the deviation of the object beam from the reference beam. If this sounds like holography to you, you are right. The photograph of the interference pattern is a hologram. Indeed, wavefronts have been reconstructed from old "interferograms." This, however, is not what we generally mean by holographic interferometry.

Classical interference patterns are easy to interpret (in principle); every fringe represents an optical path difference of $\lambda/2$. The trouble is that the reference beam was always more or less spherical (or planar in the limit). This is very convenient for almost spherical object waves, but it made interferometric studies of general, three-dimensional objects impossibly complicated. What was needed was a way to simplify the interference pattern by matching the reference beam closely to the object beam. Indeed, if the two could be made identical, no interference pattern would exist (the "null" condition) until the object was perturbed. The perturbations would then be interpretable in terms of classical theory (one fringe $\leftrightarrow$ shift of $\lambda/2$).

Holography offers not just one, but three, ways to accomplish this. These three ways are called single-exposure, double-exposure, and continuous-exposure hologram interferometry.

B. SINGLE-EXPOSURE HOLOGRAM INTERFEROMETRY

In single-exposure hologram interferometry, a hologram is made of the object, and the reconstructed virtual image is made to coincide with the still-illuminated object. If the reconstruction is done properly, the null condition results. Therefore any change in the object produces fringes in real time.

It is usually adequate for analysis to assume that the changes in the object wavefront are small and, hence, affect the phase more than the amplitude. Thus, if the reconstructed (original) object wavefront at (x, y, z) is

$$O(x, y, z) = A(x, y, z) \exp [i\phi(x, y, z)], \tag{XII-1}$$

we can represent the perturbed wavefront by

$$O'(x, y, z) \cong A(x, y, z) \exp [i\phi'(x, y, z)]. \tag{XII-2}$$

Thus, we expect bright fringes wherever

$$|\phi'(x, y, z) - \phi(x, y, z)| = 2\pi m, \qquad m = 0, 1, 2, \ldots. \tag{XII-3}$$

If we focus on the object, we see it crossed by fringes which can be interpreted in terms of object perturbations using Eq. XII-3. In this form, single-exposure holography is very useful in searching for characteristic object deformations. Of course there is no reason to exclude continuous motion of the object. In this case the fringes are time dependent, which provides for real-time observation of changes in the object.

It is important to note that the perturbations need not be small to produce fringes. If the perturbations are large, simple interpretations may be difficult. In this form single-exposure holographic interferometry highlights regions where changes occur.

Of all the forms of holographic interferometry, the single-exposure form is the most difficult to accomplish because realignment of the developed hologram is a very critical operation. The problem is especially bad for a diffuse object beam. Several solutions have been suggested: the hologram can be developed in place to avoid the problem[7,8]; special configurations can be used which simplify the realignment problem[9,10]; and, perhaps the best solution, careful design of the plate holder can allow precision replacement and adjustment.

C. DOUBLE-EXPOSURE HOLOGRAM INTERFEROMETRY

In double-exposure hologram interferometry, two object wavefronts (occurring sequentially in time) are recorded and "played back" simultaneously. Figure XII-1 shows a double-exposure hologram of a plastic bottle before and after it was pressurized with air.

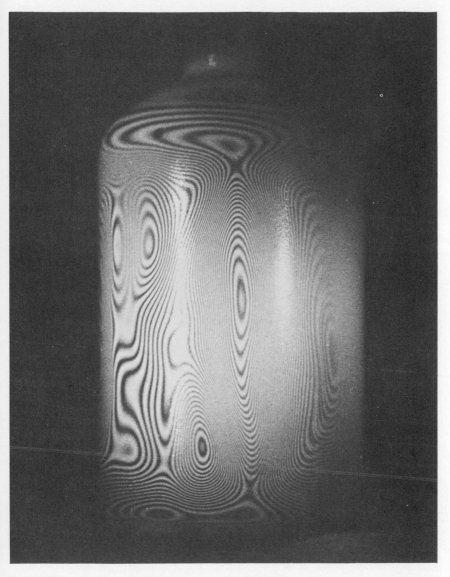

Figure XII-1 Double exposure holographic interferometry of a plastic bottle before and after it was compressed by a vice.

It is clear that the double-exposure method gives a comparison between two states of the object in precisely the same way as does the single-exposure method, however, the double-exposure method is easier because it avoids the realignment problem. On the other hand, the double-exposure method can only compare the object and one perturbed state. Thus, the single-exposure method is more versatile.

Of course, multiple-exposure holograms[11] can be produced in the same manner, with each exposure representing an incremental change in the object. This method produces very sharp fringes and, hence, leads to accurate quantitative results.

D. CONTINUOUS-EXPOSURE HOLOGRAM INTERFEROMETRY

In continuous-exposure hologram interferometry, the object moves continuously during the exposure time. There are a number of ways to describe the resulting pattern, three of which are particularly instructive. According to one explanation, that of the originators of this method, Powell and Stetson,[1] the hologram can be thought of as the limiting case (as $N \to \infty$) of an N-exposure hologram. That is, the image is the wavefront

$$U = \lim_{N \to \infty} \frac{1}{t_e} \sum_{n=1}^{N} U_n t_n, \qquad \text{(XII-14)}$$

where U_n is the wavefront at some time between $\sum_{m=1}^{n} t_m$ and $\sum_{m=1}^{n+1} t_m$: $t_e = \sum_{n=1}^{N} t_n$; and the t_n's all approach zero as $N \to \infty$. For the moment we consider only a planar object with a small vibration δz. The object wavefront at time t_n is

$$U_n(x, y, z, t) = U_0(x, y, z) + \delta U(t_n). \qquad \text{(XII-5)}$$

If the motions are very small and are sinosoidal, U_0 is the average wavefront and $\delta U(t_n)$ can be thought of as a displacement of the whole U_0 pattern to a position $x + \delta x$, $y + \delta y$, $z + \delta z$. Clearly the δz term dominates in vibration, so

$$U_n \cong U_0[x, y, z + \delta z(t_n)]. \qquad \text{(XII-6)}$$

Finally, if δz is small enough, U_0 does not change shape so much as it changes phase. From Figure XII-2, it is easily shown that

$$U_n = U_0(x, y)e^{i(2\pi/\lambda)\delta z(\cos \alpha + \cos \beta)}, \qquad \text{(XII-7)}$$

where $\alpha(\beta)$ is the angle between the viewing (illuminating) direction and the z axis. Finally, we can write the z motion as

$$\delta z(t) = m(x, y) \cos [\omega t + \theta(x, y)], \qquad \text{(XII-8)}$$

where ω is the frequency of vibration. It can then be shown[1,12] that the reconstructed image is

$$U_I \cong J_0 \left[\frac{2\pi}{\lambda} (\cos \alpha + \cos \beta)m(x, y) \right] x U_0(x, y, z), \qquad \text{(XII-9)}$$

where J_0 is the zero-order Bessel function. The fringes are contours of constant amplitude. The bright areas occur at the nulls of the vibration pattern. The same expression can be derived from considering the blurring of the interference pattern due to the motion of various parts of the object. This also shows quite clearly why the bright areas correspond to nulls. Perhaps the best intuitive notion of the patterns can be gained from the highly simplified

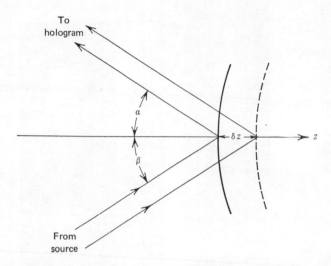

Figure XII-2 Calculation of phase change during vibratory motion based on an analysis by Monahan and Bromley.[12]

model of Monahan and Bromley,[12] in which the continuous-exposure hologram is viewed as being much like a double-exposure hologram in which the two exposures represent the positions where the most time is spent, that is, where the speed of vibration is zero—the extrema. The predictions on the basis of this simple theory agree with the exact theory given above and lend physical insight into the equations. Figure XII-3 shows a series of continuous-exposure holograms of an x-cut quartz plate parametrically excited in various modes.

Several variations of this method are of interest. It can be combined with stroboscopic techniques by using an electrooptic modulator synchronized with the vibrating object.[13,14] The timing is set so that this amounts to double- or multiple- exposure during each vibration cycle. This is much like the pulsed double-exposure method.[2] For many purposes it is perfectly adequate to use the continuous-exposure method for nonperiodic motion. No general theory is possible without specifying the nature of the motion, so quantitative interpretation may be difficult.

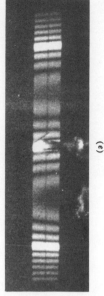

(a)

(b)

(a)

(b)

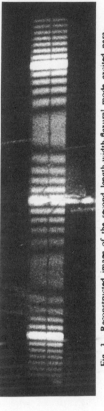

Fig. 1. Reconstructed images of the directly excited fundamental length-extensional mode of an X-cut plate, showing amplitude distribution of the vibration displacement along (a) the width direction and (b) the length direction. (Plate dimensions are 25.0 by 2.50 by 0.50 mm.)

Fig. 2. Reconstructed images of the directly excited second length-width-flexural mode of an X-cut plate, showing amplitude distribution of the vibration displacement along (a) the width direction and (b) the length direction. (Plate dimensions are 25.0 by 2.50 by 0.50 mm.)

Fig. 3. Reconstructed image of the second length-width-flexural mode excited parametrically by the fundamental length-extensional mode of a 5 X-cut plate. (Plate dimensions are 40.0 by 3.78 by 1.00 mm.)

Figure XII-3 A series of continuous-exposure holographic interferograms of X-cut quartz plates. Courtesy of Y. Tsuzuki, Y. Hirose, and K. Iijima, *Proc. IEEE,* **56,** 1229 (1968).

E. APPLICATION AREAS

The most obvious application areas are those in which ordinary interferometry is used. Smith[15] has described several such applications, for example, a holographic version of a Twyman-Green interferometer to measure imperfections in optical flats and interferometry through intervening media of poor optical quality. In these cases new flexibility is added by holography.

Holographic interferometry is compatible with many of the other techniques of holography, so the combination of interferometry with such other techniques as microscopy, pulsed holography, and "fog penetration" offers a powerful tool in special applications.

Holograms generated nonoptically can be used, in principle at least. This opens the possibility of electron interferometry, ultrasonic interferometry, comparison of real objects with their computer-generated holographic images, etc.

One of the unique applications of holographic interferometry is the detection of any change or deformation of an object during a long time period. For example, the "initial state" of an object can be recorded permanently on a hologram; at any later time (years if desirable), the object can be compared interferometrically with the reconstructed image of its initial state. The interferogram obtained will reveal any change of the object during the time span.

Another important application of the technique has been demonstrated by Archbold et al.[16] The shape of a precision mechanical object is recorded on a hologram. This stored master image is then used to make interferometric comparisons with each newly produced piece of the same object. The interference pattern obtained shows the degree of precision of the production. A very promising use of this particular technique is to make highly accurate inspection of optical components. For example, the quality of the reproduced aspherical lenses can be inspected accurately in a glance at the interferogram obtained with a stored master image of a standard piece.

F. COMPARISON WITH CLASSICAL INTERFEROMETRY

Let us consider the same experiment performed classically and holographically with the setup shown in Figure XII-4. A sample is placed in a test chamber, the wavefronts from the two paths are superimposed, and the interference pattern is interpreted as the result of distortions introduced by the sample. In order for that interpretation to be classically correct, there must be no distortions due to the beam splitters, the lens, or the windows of the test chamber. With holography, however, the only requirement is that the distortions introduced by those optical components not change when the sample is introduced. Thus, no precision alignment of components, no optical flats, and no high quality optical components are required. Interferometry through a dirty bell jar is no more difficult than interferometry through a vacuum. Furthermore, if the object wavefront is too complicated, the resulting interferogram with the plane wavefront will have so many

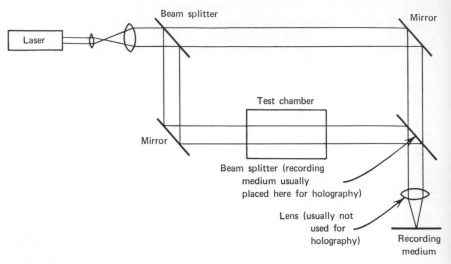

Figure XII-4 A typical interferometry setup that can be used for either classical or holographic interferometry.

complicated, closely spaced fringes that interpretation will be impossible. On the other hand, holographic interferometry does compare the diffuse object wavefront, not to a plane wave, but to a closely matched diffuse wavefront. Thus, holographic interferometry allows the use of diffuse object wavefronts for the first time.

We have seen two unique types of advantages of holographic interferometry over classical interferometry. First, it opens up new possibilities (interferometry through imperfect media, interferometry with diffusing objects, etc.) and, second, even when classical interferometry is possible, holographic interferometry is easier (less expensive equipment, less critical alignment, etc.).

REFERENCES

1. R. L. Powell and K. A. Stetson, *J. Opt. Soc. Amer.*, **55**, 1953 (1965).
2. R. E. Brooks, L. O. Heflinger, and R. F. Wuerker, *Appl. Phys. Letters*, **7**, 248 (1965).
3. K. A. Haines and B. P. Hildebrand, *Phys. Letters*, **19**, 10 (1965).

4. J. M. Burch, *Prod. Eng.* **44**, 431 (1965).

5. R. J. Collier, E. T. Doherty, and K. S. Pennington, *Appl. Phys. Letters*, **7**, 223 (1965).

6. M. H. Horman, *J. Opt. Soc. Amer.*, **55**, 615 (1965).

7. J. D. Bolstad, *Appl. Opt.*, **6**, 170 (1967).

8. D. H. Casler and H. D. Pruett, *Appl. Phys. Letters*, **10**, 341 (1967).

9. G. B. Brandt, *Appl. Opt.* **6**, 1535 (1967).

10. L. J. Tanner, *J. Sci. Instr.* **43**, 81 (1966); *Appl. Opt.* **7**, 987 (1968).

11. J. Pastor, in *Applications of Lasers to Photography and Information Handling*, R. D. Murray, Ed., Society Photographic Scientists and Engineers, Washington, 1968, Chapter 5.

12. M. A. Monahan and K. Bromley, "Vibration Analysis by Holographic and Conventional Interferometry," U. S. Government Report AD 663271, 1967.

13. G. E. Archbold and A. E. Ennos, *Nature*, **217**, 942 (1968).

14. P. Shajenko and C. D. Johnson, *Appl. Phys. Letters*, **13**, 44 (1968).

15. H. M. Smith, *Principles of Holography* Wiley, New York, 1969.

16. G. E. Archbold, J. M. Burch, and A. E. Ennos, *J. Sci. Instr.*, **44**, 489 (1967).

XIII

Applications to Holographic Information Storage

A. TYPES OF STORAGE PROBLEMS

There are two types of storage problems which can be approached by holography. Both problems deal with storing as much information as possible in a certain area and reproducing the information in a useful form. In one case, the object is merely to reproduce the information. We will call this "total-access storage." In the other case, the object is to reproduce a randomly selected part of the information without reproducing the other parts. We will call this "selective-access storage." The two are clearly related. The selective-access storage capacity cannot be greater than the total-access storage capacity, for the former is a special case of the latter.

B. TOTAL-ACCESS STORAGE

Here we raise the question: "Will holography surpass microfilm and similar media?" The answer is, "yes." This can be demonstrated as follows. Consider the best conventional storage medium. It can be assumed to have M resolvable cells/cm^2 and to be capable of N bits of intensity per cell. We define the number of bits per cell as the logarithm to the base 2 of the maximum transmission range divided by the average transmission fluctuation. The information storage capacity of ordinary photography using this medium is NM bits/cm^2. For example, we can evaluate Kodak 649F plates. Experimentally we know $N \approx 1$ bit and $M \approx 4 \times 10^8$ cells/cm^2 or $NM \approx 4 \times 10^8$ bits/cm^2. We know that holography using that medium can reconstruct an image with M cells/cm^2 as discussed in Section VI-E. The question is: "Can the information per cm^2 in the image plane exceed NM bits?" Surprisingly, the answer is, "yes." This is not a violation of the second law of thermodynamics. Rather it stems from the different way the information is read out in the two cases.

Each resolution cell in the recording medium records the signal at that location with additive noise. If each resolution cell in the image receives contributions from C resolution cells in the hologram, this noise can be effectively canceled out. This is closely related to coding theory; and Shannon's coding theorem suggests that for a given recording medium, we should be able to achieve arbitrarily good noise cancellation so long as C is greater than some finite number, C_0, called the "channel capacity." Other factors which contribute to noise in the image are scattering from dirt, scratches and other imperfections, and misdirected signal due to distortions of the hologram. With proper care, very high signal-to-noise ratios can be achieved.

Now let us summarize the argument. Given a square centimeter of film, we can achieve equal numbers of resolution cells in the image using either holography or photography, but we can pack in more information per cell with holography. This is illustrated by Kodak 649F plates, for which $N \approx 1$ for ordinary photography, but $N \approx 20$ has been achieved by holography.

There are two great difficulties with utilization of this great number of "shades-of-gray." First, most optical information is not in this form, and, second, it is not easy to read out that much information. The latter problem is strictly technical and can be solved. The former problem is more fundamental. The human eye, is not a very good instrument for making absolute intensity judgments. For this reason we encode our most important information spatially. The page you are now reading contains only binary intensity information ($N = 1$). Most of the information is in terms of shapes and locations of those shapes. Thus, we must encode the usual optical information in some way to allow better utilization of holography. All codes involve waste of information capacity, so holography will never be fully utilized for data storage.

Besides increased storage capacity, holography offers other significant advantages over ordinary photography for data storage. First, the possibility of distributing the information spatially provides a large amount of immunity to information loss from damage to the hologram. Second, holography is essentially a projection technique, so read out of the information involves no contact with (or damage to) the hologram.

C. SELECTIVE-ACCESS STORAGE

The idea of selective-access holographic information storage is to record a large number of images (messages) on a single hologram in such a way that they can be reconstructed one at a time. There are a number of ways of doing this. The images can be stored in physically separated parts of the hologram, or they can be distributed throughout the hologram. If the images are stored throughout the hologram they must be separately addressable. Since addressing is done by the reference beam, the reference beam must be different for each image. The types of distributed storage techniques are distinguishable by the ways in which the reference beam can be varied.

First we will discuss physically separated storage. We noted in Section VI-C that any part of a hologram taken with a diffused object beam reconstructs a view of the object. If an opaque screen with a small hole in it is placed in front of the recording medium prior to exposure, the hologram will exist only in that localized spot. Before we examine the information storage capacity characteristic of this method, we will examine some characteristics of the reconstructed image. Since the entire information capacity of any one part of the hologram is used to reconstruct one image, the maximum signal-to-noise ratio can be obtained for that image. The bandwidth-reduction technique of Haines and Brumm[1] can be used to achieve perspective characteristic of much larger apertures. The number of images which can be stored is the two-dimensional space–bandwidth product of the recording medium divided by the space–bandwidth product needed to record a single image. For a plate of dimensions $L_1 \times L_2$ and resolution length l, the space–bandwidth product is

$$(SW)_{\text{Plate}} = \frac{L_1 L_2}{l^2}. \tag{XIII-1}$$

For an object of dimensions $D_1 \times D_2$ and resolution length d_2, we have

$$(SW)_{\text{Object}} = \frac{D_1 D_2}{d_2^2}. \tag{XIII-2}$$

Finally, from Lohmann's analysis of split-beam Fresnel holography, we have

$$(SW)_{\text{Hologram}} \approx 64(SW)_{\text{Object}}. \tag{XIII-3}$$

Therefore, we should be able to store no more than

$$n_{\text{max}} \approx \frac{1}{64}\left(\frac{L_1 L_2}{D_1 D_2}\right)\left(\frac{d_2}{l}\right)^2 \tag{XIII-4}$$

holograms. Consider the case of Kodak 649F plates. $L_1 = 4$ in. $L_2 = 5$ in., and $l \approx 1/2200$ mm $\approx 1.8 \times 10^{-5}$ in. As a measure of a useful image bandwidth, let us choose the 525-line television. So

$$\frac{D_1 D_2}{d^2} = (525)^2 \approx 2.75 \times 10^5. \tag{XIII-5}$$

Then

$$n_{\text{max}} \approx 1.4 \times 10^3. \tag{XIII-6}$$

Naturally we will not be able to achieve the full n_{max}, but large numbers of TV quality images can be stored in this way.

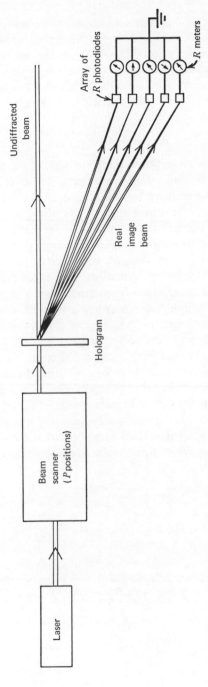

Figure XIII-1 A system for storage and retrieval of digital information using holography. A light scanner positions a laser beam on one of P distinct holograms on the recording medium. Each hologram causes a real image of up to R points of light to fall on R photodetectors. Thus any one of PR spots can be read out.

121

The selectivity of thick holograms with respect to wavelength and direction was described in Section V-D. It can be shown[2] that the number of images that can be stored in a three-dimensional medium is the number of resolution elements in the medium divided by the space–bandwidth product of the image. For a 1-cm^3 medium with resolution length $l = 1/2200$ mm and an image space–bandwidth product, again, $(525)^2$, the number of images which can be stored is

$$n_{max} \approx 4 \times 10^7.$$

Clearly, this is potentially the better storage method. The achievement of this type of storage capacity depends on the development of thick, high resolution media, such as photochromic glass[3] or bleachable alkali halide crystals.[4]

These same techniques also suffice to allow an optical associative memory[2] wherein a unique stored image is reconstructed for each of a fixed set of input images. Naturally cross-talk will hold the number of associative image pairs to well below the maximum possible in principle.

For very fast access to the holographically stored information, various non-mechanical light beam scanners have been suggested. The basic scheme[5] is shown in Fig. XIII-1. The light beam scanner is used to illuminate one of P "pages." Each page is a hologram of an array of R point light spots. Each of the R spots falls on one of R photodetectors which read the page. Thus, a total of $P \times R$ points can be read from a single hologram. Present projections indicate that $P \times R$ values up to 10^{12} may be obtained by the mid-1970's. A great increase in the number of electronically accessible points can be gained by using the scanner to address Q holograms in each of its P beam positions and controlling which of the Q holograms actually projects the R spots. Studies are now under way in several locations to determine if there are feasible, nonmechanical ways of doing this. The total number of accessible spots is then $P \times R \times Q$.

REFERENCES

1. K. A. Haines and D. B. Brumm, *Appl. Opt.*, **7**, 1185 (1968).

2. P. J. van Heerden, *Appl. Opt.*, **2**, 387 (1963).

3. D. R. Bosomworth and H. J. Gerritsen, *Appl. Opt.*, **7**, 95 (1968).

4. G. U. Kalman, in *Applications of Lasers to Photography and Information Handling*, R. D. Murray, Ed., Society Photographic Scientists and Engineers, Washington, 1968, pp. 99–104.

5. F. M. Smits and L. E. Gallaher, *Bell Syst. Tech. J.*, **46**, 1267 (1967).

XIV

Applications of Holography to Microscopy

A. INTRODUCTION

Two forms of holographic microscopy must be distinguished. One form uses holography in conjunction with ordinary microscopy. This form can be called "holographically augmented microscopy." The other form uses the hologram properties themselves to do the magnifying. This form can be called "holographic magnification." The two forms of holographic microscopy are fully compatible.

B. HOLOGRAPHICALLY AUGMENTED MICROSCOPY

Holography can augment normal optical microscopy in any of several ways. The hologram can furnish the image for ordinary microscopy. The image wavefront formed by the microscope objective can be recorded holographically for subsequent viewing with the eyepiece. The image wavefront viewable through the eyepiece can be recorded on a hologram. The holographic tricks of interferometry (Chapter XII), contour formation (Section VIII-I), polarization recording (Section VIII-D), color recording (Section VIII-D), etc. can all be employed to augment ordinary microscopy. Among the pioneers in this field were Van Ligten and Ostenberg,[1] Carter and Dougal,[2] and Ellis.[3]

The single greatest advantage of holographically augmented microscopy is the ability to use pulsed lasers to record the object information in a short time and in full three dimensions. Thus, transient objects can be examined at leisure and in depth with a variety of techniques.

A remark must be made concerning the three dimensionality of the image. There is a fundamental limit on all imaging systems (holographic or ordinary) relating to focus properties and focal numbers. The focal number N is the ratio of the focal distance to the diameter of the aperture. The focus is distinct for lateral dimensions greater than a minimum length

$$l_L \propto N^{-1}.$$

(XIV-1)

The focus in the depth dimension is distinct only for a depth less than a depth of field

$$l_D \propto N^2. \tag{XIV-2}$$

Briefly, this means that if high resolution (low l_L) is to be achieved, then very small depth of focus (small l_D) will result. Thus, at any one time there can only be a region of about $l_L \times l_L \times l_D$ dimensions in clear focus. For microscopes l_D is so small that the three dimensionality of the image is negligible. The holographic image is three-dimensional only in the sense that the focus can be adjusted to bring new parts of the image into focus.

C. HOLOGRAPHIC MAGNIFICATION

There are two basic forms of holographic magnification in use—geometric change and wavelength change.

Geometric change magnification occurs when the reconstruction geometry is changed from the construction geometry in the proper way. The theory of geometric change magnification has been summarized very thoroughly by De Veltis and Reynolds.[4] Although beautiful images have been produced with magnifications up to 120 × by this method.[5] the method is, as yet, of little practical value. The reason for this situation is easy to explain. Holograms record direction information in the form of a diffraction grating. In the process of developing the hologram, a certain absolute amount of error is inevitable in the grating. A given absolute error in the grating produces a greater angular error for a large diffraction angle than for a small diffraction angle. The diffraction angle must be large (small focal number) for high resolution; so high resolution holograms are eventually limited, not by fundamental limitations, but by noise which blurs all high resolution information.

Wavelength change without a corresponding scaling up (or down) of the hologram leads to magnification. Again the problem is treated in detail by De Velis and Reynolds.[4] In the most general case geometric changes and wavelength changes may occur simultaneously, so no general formula for magnification exists. In many formulas the magnification increases as the ratio of the reconstructing wavelength, λ_R, to the constructing wavelength, λ_C, increases.

The problem of achieving $\lambda_R/\lambda_C \gg 1$ cannot be solved with λ_C in the visible region since the reconstruction wavelength must be visible (for viewing purposes), and the largest λ_R/λ_C for both wavelengths in the visible region is $\lambda_R/\lambda_C \approx 2$. Thus, we must construct the hologram with $\lambda_C \ll 0.4\mu$ (violet light). Proposals have been made for the use of electron waves[6-9] and X-rays.[6,10-14] To date, neither technique has been very successful, although the future (Chapter XVI) may hold more promise in this area.

It is important to remember that in holography (as in other imaging systems) magnification leads to distortion. The lateral magnification, m_{lat}, is related to the longitudinal magnification, m_{long}, by

$$m_{long} = \frac{\lambda_C}{\lambda_R} m_{lat}^2. \tag{XIV-3}$$

D. THREE-DIMENSIONAL RECONSTRUCTION

We noted above that high magnification led to restrictive field of view and limited depth of field. To date, only two ways of "solving" this problem have been demonstrated. Knox[15] recorded the unmagnified object and examined the reconstructed image with a microscope. Toth and Collins[16] recorded the object through a magnifying lens. Then the real image is reconstructed in such a way that the light passes back through the lens to form a three-dimensional, unmagnified image which is essentially aberration-free and has a much larger observable volume than would be possible with a hologram of the magnified image. The method of Knox provides more information than the method of Toth and Collins, but the latter requires less film resolution and, hence, may find wide application.

E. APPLICATIONS

We mentioned in Section IV-C that the holographic magnification device of Thompson et al. was the first commercially exploited application of holography. It has been used to study particles, aerosols, etc. The most recent review of this work was given by Ward.[17]

Hologram microscopy is of most value when the object is three dimensional and transient or moving. In this case it offers the only real solution.

Coherent light can be used for various optical image processing techniques, and holography is the only way to record and reconstruct all the object information necessary for this type of examination. An example of this approach is the use of Hilbert transforms to observe phase objects.[18]

Of course, all the "tricks" of holography (see especially Chapter VIII) can be applied to the microscopic case when they seem to offer advantages.

REFERENCES

1. R. F. van Ligten and H. Osterberg, *Nature*, **211**, 282 (1966).
2. W. H. Carter and A. A. Dougal, *IEEE J. Quantum Electron.*, **2**, 44 (1966).
3. G. W. Ellis, *Science*, **9**, 1195 (1966).
4. J. B. De Velis and G. O. Reynolds, *The Theory and Applications of Holography*, Addison-Wesley, Reading, Mass., 1967.

5. E. N. Leith and J. Upatnieks, *J. Opt. Soc. Amer.*, **55**, 569 (1965).

6. D. Gabor, *Nature*, **161**, 777 (1948).

7. M. E. Haine and J. Dyson, *Nature*, **166**, 315 (1950).

8. E. J. Thompson, *Japan. J. Appl. Phys.*, **4**, Supplement **1**, 302 (1965).

9. A. Tonomura, A. Fukuhara, H. Watanabe, and T. Komoda, *Japan. J. Appl. Phys.*, **7** 295 (1968).

10. M. J. Buerger, *J. Appl. Phys.*, **21**, 909 (1950).

11. A. V. Baez, *Nature*, **169**, 963 (1952).

12. H. M. A. El-Sum, and A. V. Baez, *Phys. Rev.*, **99**, 624 (1955).

13. J. T. Winthrop and C. R. Worthington, *Phys. Letters*, **15**, 124 (1965).

14. G. W. Stroke and D. G. Falconer, *Phys. Letters*, **13**, 306 (1964).

15. C. Knox, *Science*, **153**, 989 (1966).

16. L. Toth and S. A. Collins, Jr., *Appl. Phys. Letters*, **13**, 7 (1968).

17. J. H. Ward, in *Applications of Lasers to Photography and Information Handling* (R. D. Murray, Ed., Society Photographic Scientists and Engineers, Washington, D.C. 1968, p. 217.

18. S. Lowenthal and Y. Belvaux, *Appl. Phys. Letters*, **11**, 49 (1967).

XV

Applications of Holography to Motion Pictures and Television

A. INTRODUCTION

Movies and television are two of our most important entertainment media. Both would be revolutionized by three-dimensional imaging.

Another important aspect of movies and television is their scientific use for data transmission and display. Here holography offers many advantages. The ability to transmit images of very high contrast is an obvious advantage. Less obvious is the fact that the resolution and contrast function for holograms are affected differently by television information losses than they are for ordinary point-by-point transmission. This will allow such phenomena as resolution exceeding the resolution of the scanned point.

Movies and television have certain things in common. They both require that a sequence of images be produced and reconstructed. The difference between successive images must reflect motion during that period, but no single image must show blurring due to motion. If we assume that the scene to be recorded contains human beings, we can draw several conclusions. In order to stop the motion we need very short pulses (Section VII-B). It seems safe to assume that the 40 nsec. time used by Siebert[1] to record humans can be relaxed slightly. Let us assume that a light pulse of 10^{-4} sec is used. If we assume that 16 frames/sec will be used, then the duty cycle is $(1/16)/10^{-4} = 625$. The energy reaching the recording medium during the 10^{-4} sec exposure must be sufficient to record the hologram. Kodak 649F film requires 10^3 erg/cm^2. Let us assume, optimistically, that we only need 10 erg/cm^2. The power density required on the film is

$$\mathscr{P} = \frac{(10 \text{ erg/cm}^2)(10^{-7} \text{ j/erg})}{10^{-4} \text{ sec}} = 10^{-2} \text{ W/cm}^2.$$

This is high compared with ordinary ambient light levels. Therefore, by using light pulses and shutters which stay open only 10^{-4} sec, we can take the movie or television holograms under normal lighting conditions without

127

filtering out the background light. The apparent brightness is associated with the time-averaged power density $(10^{-2}/625) = 1.6 \times 10^{-5}$ W/cm^2. Unfortunately, the power density on the object must be orders of magnitude greater than this, so eye damage is a possibility. The possibilty of eye damage stems from the possibility that the human will focus on the point laser source, thus forming a point image on the retina which is powerful enough to damage the sensors. The brightness of image is independent of the distance between the observer and the object. One way to aleviate this problem is to illuminate the scene via an extended, diffusing disc.

B. MOVIES

The main difficulty in holographic movies is taking the holograms. Holograms produced on film rather than on plates reconstruct very well. It would be very easy to modify a standard movie projector (by inserting a diaphragm spatial filter and some sort of color filter) so that it would reconstruct hologram movies. Alternatively, the holographic information storage techniques can be used to place the whole movie on one hologram.

One particularly interesting approach to holographic movies, that of De Bitetto,[2] avoids the necessity of discontinuous film motion. De Bitetto notes that vertical parallax is seldom of value (human eyes being separated horizontally), so it can be virtually eliminated without serious loss is the "three dimensionality" of the scene. He then shows that horizontal-strip holograms[3] which provide little vertical parallax can be moved vertically without much image degradation. A series of strip holograms moved vertically will produce a series of images with good horizontal parallax.

A second difficulty with holographic movies arises if one wishes to have the movie viewable by a large audience. The angular field of view of most holograms is limited by the size of the hologram. The object can be viewed only from those directions from which we could view the original object through a window the size and location of the hologram. There are only two choices, neither of them encouraging. Either we must record holograms on media the size of movie screens, or we must find a new way of viewing the holographic movie. The use of a large encoding screen to record a wide angular field of view on a small hologram (already suggested by Haines and Brumm[4] for an entirely different purpose) may be of advantage here also.

C. TELEVISION

Television offers not only the recording problem, but also a transmission problem and a reconstruction problem, none of which is easy to solve. The state of the art in 1965 was summarized by Leith et al.[5] Their paper pointed out the major obstacles then preventing holographic television. This inspired numerous efforts to solve these problems. While great progress has been made, holographic television is still impractical at the time of this writing.

The transmission problem is easy to describe. The system must have a time–bandwidth product in excess of the three-dimensional space–bandwidth product of the scene. Leith et al.[5] show that the scene space-bandwidth product is about 4.9×10^9. At 30 frames/sec., this requires broadcasting at a rate of about 1.5×10^{11} samples/sec. compared with about 8.4×10^6 samples/ sec. for normal television. Since that time numerous techniques to reduce the bandwidth for transmission have been suggested.[3-6] Let us be perfectly frank about how these bandwidth reductions occur. Part, but not much, of the reduction comes in more efficient use of the available space-bandwidth. If the number of samples is reduced below the space-bandwidth product of the scene, information is lost. Fortunately, we do not need the full 4.9×10^9 samples. For ordinary television, the two-dimensional space-bandwidth product 525×525 is perfectly adequate. The depth dimension needs even less information, so, in principle, a good image could be obtained by $525 \times 525 \times 20 = 5.5 \times 10^6$ samples or at a rate of about 1.65×10^8 samples/sec. The point is very simple. We can't get out more information than we put in. To get three-dimensional images, we must pay the price in terms of bandwidth. Even 10^8 samples/sec requires techniques beyond the state of the television art. Leith et al.[5] have suggested that light beam transmission may be the solution there.

The receiver is another problem for holographic television. First, it must be able to handle as much as 10^8 samples/sec. Second, it must display about 5.5×10^6 samples simultaneously (note that a hologram cannot be reconstructed point by point as an ordinary image can). Third, the information must modify the transmitted light by a significant amount. None of these tasks is easy. The two approaches suggested by Leith et al.[5] are an electron-beam addressed phase medium ("Eidophor"[8]) and a laser-beam scanned photochromic medium.

Clearly holographic television is possible, although many technical advances must be made before it becomes practical as an entertainment medium; however, scientific and technical uses are approaching real practicality. From the first holographic television demonstration by Enloe et al.[7] to the excellent results of Gurevich et al.,[8] much progress has been made. It is now clear that special-purpose television receivers for holography will be used for data transmission in the near future.

REFERENCES

1. L. D. Siebert, *Appl. Phys. Letters*, **11**, 326 (1967).
2. D. J. De Bitetto, *Appl. Phys. Letters*, **12**, 295 (1968).

3. D. J. De Bitetto, *Appl. Phys. Letters*, **12**, 176 (1968).

4. K. A. Haines and D. B. Brumm, *Appl. Opt.*, **7**, 1185 (1968).

5. E. N. Leith, J. Upatnieks, B. P. Hildebrand, and K. Haines, *J. Soc. Motion Picture Television Eng.*, **74**, 893 (1965).

6. L. H. Lin, *Appl. Opt.*, **7**, 545 (1968).

7. L. H. Enloe, J. A. Murphy, and C. B. Rubinstein, *Bell Syst. Tech. J.*, **45**, 335 (1966).

8. S. B. Gurevich, G. A. Gavrilov, A. B. Konstantinov, V. B. Konstantinov, Yu. I. Ostrouskii, and D. F. Chernykh, *Sov. Phys.—Tech. Phys.*, (*English Transl.*) **13**, 378 (1968).

XVI

The Future of Holography

A. PROBABLES AND POSSIBLES

We will not abuse our readers with predictions of what may be possible. Rather, we will describe a few applications of holography which are known to be possible and which are being pursued actively in laboratories throughout the world. Furthermore, these applications have such great potential that work on them is likely to continue until the solutions are obtained.

B. X-RAY HOLOGRAPHY

Let us first imagine what X-ray holography might be able to accomplish. Vastly enlarged, three-dimensional images of molecules could be produced. This would represent an unparalleled breakthrough in the physical investigation of life. Biology and medicine would be changed overnight. We would make great strides in our understanding of crystals. Cancers could be pinpointed accurately in three dimensions. X-ray therapy could become much safer and more effective by focusing in a well defined location.

There are, of course, two problems to be solved. First, we must produce a source of highly coherent X-ray radiation. Work toward the solution of the problem is under way.[1] The final solution may be many years away. Second, we must produce a recording medium capable of recording an X-ray hologram. The development of lensless Fourier transform holography (Section VIF) was the first significant step toward making the resolution requirements less severe. It can be shown [2,3] that the finite source size now inherent in X-ray sources will limit the resolution of the hologram. Fortunately, this loss is not irretrievable. Stroke et al.[4] showed that if the reconstruction occurs using a suitably structured source instead of a point source, a deconvolution restores the original image. This is another application of the general principle also used to retrieve information lost by a poor lenses (Secion VIIE) or an intervening random medium (Section VIII-F).

The first experiments on X-ray microscopy were reported by El-Sum and Baez[5] in 1955. They encountered many difficulties in the form of poor source coherence and absorption by the object. Some progress has been made since then. The latest effort has been on long-wavelength X-ray holograms.[6]

On the other hand, X-ray diffraction patterns are easy to produce. Such a diffraction pattern is a self-encoded hologram in the sense that a part of the original wavefront can be used to reconstruct the rest of the wavefront.[4] There remains some hope[7] that these holograms can be made to reconstruct the object if the resulting Patterson pattern can be deconvolved.[8]

The rewards for success are so great that effort in this field will certainly continue until the solutions are found.

C. CO_2 LASER HOLOGRAPHY

The use of CO_2 laser light at 10.6μ would offer many advantages. It would produce an automatic means for bandwidth reduction (Section XV-C). A $10.6\text{-}\mu$ wavelength hologram records only $(0.633/10.6)^3 \approx 2.2 \times 10^{-4}$ of the bandwidth of a hologram recorded with a 0.633 He–Ne laser. A CO_2 laser of essentially unlimited power can be made. This will allow holography of objects at great distances (the 10.6μ is in an "atmospheric window" and thus suffers little absorption in the atmosphere).

The normal CO_2 laser is ill-suited to holography, but techniques to make it highly coherent are available. There is no film for recording $10.6\text{-}\mu$ radiation, but a number of devices exist which produce an image from CO_2 light. Unfortunately (for purposes of this book), those devices are militarily classified, as are their performance criteria. The first CO_2 laser hologram was taken in 1968 by Lowenthal et al.[9]

D. ELECTRON MICROSCOPY

Even without the improved imaging Gabor[10] sought to provide for it, electron microscopy has become one of the most important tools of present day science. In only 20 years, electron holography caught up with normal electron microscopy, as an imaging method. The breakthrough was due to very careful and clever work by Tonumura et al.[11] To date, considerably less effort has been expended on electron holography than on electron microscopy, yet they give equal resolution. If no further improvements in either were forthcoming, holography would be preferred for some applications. It seems almost certain that great improvements in electron holography will arise from the current world wide research in all fields of holography.

E. SAMPLING

Classical sampling theory treats the problem of perfect reconstruction of a function $f(x)$ from sampled values $f(x_n)$. Holography is concerned with producing approximations to $f(x)$ by Fourier transforming a sampled wavefront which is based on samples $F(u_n)$ of the far-field pattern (inverse Fourier transform) of $f(x)$. The former was not derived with the latter in mind. Around the world the problem of sampling is being examined from the holographic point of view. The results are not firmly established as yet, but the trends are obvious. First, ways will be sought to take *a priori* knowledge of the wavefront (which is considerable for all physical objects) into account.[12] Second, optimum distributions of the sampling points are being sought.[13] Third, nonlinear processing of the data to make better use of the available information is being considered. All these innovations are aimed at improving the image which can be obtained from a given number of samples. The present indication is that the improvements will be dramatic and have profound influences in image storage, hologram transmission, nonoptical holography, etc.

F. SPECTROSCOPY

The venerable field of spectroscopy is being invaded by holography from two directions. First, holographic diffraction gratings using dichromated gelatin can be produced cheaply and accurately. The resolving power and noise are equivalent to ruled gratings, and the diffraction efficiency can exceed 90%. Second, spectroscopy without gratings is possible holographically. These techniques cause each line of the spectrum to form its own holographic diffraction grating. There are several basic approaches to this. Fourier transform holographic spectroscopy was first proposed by Stroke and Funkhouser[14] and was reviewed recently by Stroke.[15] Saccocio[16] suggested that Lloyd's mirror be used to generate a hologram of the spectrum. The experimental work on holographic spectra using Lloyd's mirror was carried out by Kamiya et al.[17] The numerous advantages which can be forseen in luminosity, brightness, resolution, and insensitivity to stray light make this an active field of research.

REFERENCES

1. R. M. Cotterill, *Appl. Phys. Letters*, **12**, 403 (1968).

2. A. V. Baez, *J. Opt. Soc. Amer.*, 42, 756 (1952).

3. C. C. Eaglesfield, *Electron. Letters*, **1**, 181 (1965).

4. G. W. Stroke, R. Restrick, A. Funkhouser, and D. Brumm, *Phys. Letters*, **18**, 274 (1965).

5. H. M. A. El-Sum and A. V. Baez, *Phys. Rev.*, **99**, 624 (1965).

6. J. W. Giles, Jr, *J. Opt. Soc. Amer.*, **59A**, 498 (1969).

7. V. V. Aristov, V. L. Broude, L. V. Koval'skii, V. K. Palyanskii, V. B. Timofeev, and V. Sh. Skekhtman, *Sov. Phys Doklady*, **12**, (*English Transl.*), **12**, 1035 (1968).

8. G. W. Stroke and D. G. Falconer, *Phys. Letters*, **13**, 306 (1964).

9. S. Lowenthal, E. Leiba, M. Lucas, and A. Werts, *Compt. Revd.*, **266B**, 1 363 (1968).

10. D. Gabor, *Nature* **161**, 777 (1948).

11. A. Tonomura, A. Fukuhara, H. Watanabe, T. Komoda, *Japan. J. Appl. Phys.*, **7**, 295 (1968).

12. H. J. Caulfield, *Proc. IEEE* **57**, 2082 (1969).

13. H. J. Caulfield, *Phys. Letters*, **28A**, 600 (1969).

14. G. W. Stroke and A. Funkhouser, *Phys. Letters*, **16**, 272 (1965).

15. G. W. Stroke, *Physica*, **33**, 253 (1967).

16. E. J. Saccocio, *J. Opt Soc. Amer.*, **57**, 966 (1967).

17. K. Kamiya, K. Yoshihara, and K. Okada, *Japan J. Appl. Phys.*, **7**, 1129 (1968).

APPENDIX: HOW TO FIND MORE OR NEWER INFORMATION ON HOLOGRAPHY

Information on holography is easy to obtain. There is so much information that the problem is to get the right information. We hope this appendix will be a useful guide.

The *Journal of the Society of Motion Picture and Television Engineers* has now published four complementary bibliographies on holography which comprise the best starting place for literature published prior to 1968. The first three are by Chambers et al.[1-3] and the fourth is by Latta.[4]

A number of review papers in holography have been published. Of these, two are of special interest. The 1965 paper by Leith and Upatnieks is still the best short introduction to holography. The 1968 paper by Brandt[6] is the latest and one of the most comprehensive.

A great deal of detailed information is available in books by Stroke,[7] De Velis and Reynolds,[8] Leith and Upatnieks,[9] and Smith.[10] Other books will be forthcoming soon.

On the particular subject of coherent optical data processing, two recent books by Papoulis[11] and by Goodman[12] are especially helpful.

The most important journals for holography are, in alphabetical order, *Applied Optics, Applied Physics Letters, Compete Rendus, Japanese Journal of Applied Physics, Journal of the Optical Society of America, Nature, Physics Letters,* and *Proceedings of the Institute of Electrical and Electronics Engineers.* The journal *Laser Focus* carries good reviews of recent developments in holography. A number of classified bibliographical journals now carry a heading "holography," although a few persist in classifying holography under headings such as "diffraction" or "physical optics."

REFERENCES

1. R. P. Chambers and B. A. Stevens, *J. Soc. Motion Picture Television Eng.*, **76**, 392 (1967).
2. R. P. Chambers and J. S. Courtney-Pratt, *J. Soc. Motion Picture Television Eng.*, **75**, 759 (1966).
3. R. P. Chambers and J. S. Courtney-Pratt, *J. Soc. Motion Picture Television Eng.*, **75**, 373 (1966).
4. J. N. Latta, *J. Soc. Motion Picture Television Eng.*, **77**, 422 (1968).
5. E. N. Leith and J. Upatnieks, *Sci. Amer.*, **212** (6), 24 (1965).
6. G. B. Brandt, *Electro-Technol.* **81** (4), 53 (1968).

7. G. W. Stroke, *An Introduction to Coherent Optics and Holography*, Academic Press, New York, 1966.

8. J. B. De Velis and G. O. Reynolds, *Theory and Applications of Holography*, Addison-Wesley, Reading, Mass., 1967.

9. E. N. Leith and J. Upatnieks, *Progr. Opt.* **6**, 100 (1967).

10. H. M. Smith, Principles of Holography, Wiley, New York, 1969.

11. A. Papoulis, *Systems and Transforms with Applications in Optics*, McGraw-Hill, New York, 1968.

Index